Praise for Evilicious

In this original and uniquely informative book, Marc Hauser shows us how the "addiction to evil" – the persistent subjection of innocents to gratuitous cruelty – emerged as a by-product of the human brain's unique evolutionary design. The ability to creatively combine all manner of thought and emotion enabled our species to produce great works of art and science, as well as to freely choose to kill and torture with a level of maliciousness unprecedented in the history of life on earth. Here we find that the most dangerous and effective evildoers are not sadists or serial killers with disordered minds, but mostly normal people who could have chosen not to kill and torture. When driven by unsatisfied desires – especially if channeled into dreams of glory for a cause – and in denying the reality and the humanity of others, even nice guys can become massively bloodthirsty.
— **Scott Atran**, *Director of Research in Anthropology at France's National Center for Scientific Research and author of "Talking to the Enemy"*

"Evilicious is an incisive and engaging analysis of why people have the capacity to inflict great evil. Marc Hauser cogently draws from psychology, neuroscience and evolution to explore potential explanations for this darkest side of human nature. His fascinating book is lively from start to finish, and helps bring science to bear on an issue of great importance."

— **Kent Berridge**, *Professor of Neurosciences, University of Michigan*

"The problem of evil is as old as recorded thought, and one might have guessed that there would be nothing fresh and original to say. But Marc Hauser's *Evilicious* is an entertaining and compassionate essay refutes that expectation, offering new perspectives and intriguing suggestions on traditional problems from a remarkably wide range of comparative and experimental evidence and evolutionary considerations, often fascinating in themselves. A thought-provoking inquiry."

— **Noam Chomsky**, *Institute Professor and Professor of Linguistics, MIT*

"The urge to explain evil is as compelling and as daunting as the wish to explain love, a task that for all of human history has befuddled poets, philosophers, psychologists, biologists and, most recently, neuroscientists. Marc Hauser draws on all those disciplines and more to examine the worst of human behavior. It is always fascinating to read what he is thinking, all the more so when he's thinking differently from me. He challenges my assumptions and makes my world bigger. Who could ask for more? Even when you disagree with him, you'll do so with pleasure."

— **Randy Cohen**, *Original writer of "The Ethicist", New York Times Magazine*

"*Evilicious* is a gripping investigation into our appetite for cruelty and destructiveness. Highly informative, and written in a lively style that is accessible to the general reader while also having much to offer the specialist, it is a book that is difficult to put down. It is a must-read for anyone interested in the puzzle of human violence, and every serious student of human nature."

— **David Livingstone Smith**, *Ph.D., Professor of Philosophy and author of "Less Than Human: Why We Demean, Enslave, and Exterminate Others."*

"In Evilicious, renowned neuroscientist Marc Hauser provides a provocative and insightful account of evil as the consequence of ancient neural systems put to new uses as evolution shaped the uniquely human, combinatorial brain. While making his case, Hauser offers lively and accessible explanations of many cutting-edge findings in social neuroscience that are a fascinating read for the general public and scientists in the field alike. The book is an impressive demonstration of the important role science plays in thinking about the most pressing social and ethical issues of today."

— **Andreas Meyer-Lindenberg**, *Director, Central Institute of Mental Health, Mannheim, Professor, University of Heidelberg*

"What Steven Pinker has done for violence, Marc Hauser has achieved with evil - this book brings the light of science to illumine the heart of darkness."

— **Nicholas Wade**, *former Science Editor,*
New York Times

"*Evilicious* offers an analysis of the human capacity for evil, and a plausible account of how it emerged, that is immediately engaging and deeply satisfying. Saturated with eye-catching examples, bristling with punchy observations, and marvelously comprehensive in its coverage, this is a book to savor and to treasure."

— **Philip Pettit**, *University Professor of Politics*
and Human Values, Princeton University, and
Distinguished Professor of Philosophy, Australian
National University.

"Marc Hauser's cogent and concise study on the psychological nature of evil could not come at a more propitious time after the Sandy Hook elementary school massacre changed the national debate on violence. Much discussion has been focused on mental health issues, but as Hauser reveals through his thorough summary of the scientific study of violence, the problem is not a handful of crazy people; it is that we all have the capacity to commit acts of violence due to the nature of our psychology and how our brains are wired. Every

Congressman, Senator, and journalist voting or writing on what to do about violence should read this book first."
— **Michael Shermer**, *Publisher of Skeptic magazine, monthly columnist for Scientific American, adjunct professor at Claremont Graduate University, and author of "The Believing Brain."*

"*Evilicious* is a serious attempt to understand gratuitous cruelty. Evolutionary psychologist Marc Hauser argues that even though the capacity for evil evolved by chance, and despite the strong roles of individuals' genes and experiences, the abuse of power makes grisly sense. Unflinching, thought-provoking and richly informed, this examination of the darkest of human desires is an important original contribution to the science of human moral failings."
—**Richard W. Wrangham**, *Professor of Human Evolutionary Biology, Harvard University, author of "Catching Fire" & "Demonic Males."*

evilicious

Marc D. Hauser

For Jacques and Bert Hauser,
my parents,
my friends,
and my reminder
that life should be
lived to its fullest,
always.

Pleasure is the greatest incentive to evil.

– Plato

To witness suffering does one good, to inflict it even more so.

– Friedrich Nietzsche

Man produces evil as a bee produces honey.

– William Golding

Dear reader,

I am familiar with the horrors of evil, mostly by indirect means. I lived in Uganda and spoke with people who witnessed the brutal savagery of the dictators Milton Obote and Idi Amin. I met child soldiers, holding automatic rifles and standing next to piles of skulls. I listened to stories of my father's childhood as a Jew hiding in Nazi-occupied France, and read countless descriptions of past- and present-day genocides. I carried out research with populations of psychopaths and bullies, while personally experiencing the bullying of individuals who had little regard or compassion for someone who couldn't fight back.

I am also familiar with and deeply moved by human kindness, particularly our capacity to reach out and help strangers. When my father was in a boarding school in the south of France, hiding from the Nazis, a little girl approached him and asked if he was Jewish. My father, conditioned by his parents to deny his background, said no. Sensing doubt in my father's voice, she replied "Well, if you *are* Jewish, you should know that the director of the school is handing Jewish children over to the Nazis." My father promptly called his parents, who immediately picked him up, thus allowing him to survive and tell me this story. This little girl expressed one of our species' signature capacities: the ability to show compassion for another person, even if that person's beliefs and desires are different from one's own.

These experiences, together with my long-held interest in and exploration of human nature, have motivated this book and my desire to explain the causes of evil. It is a topic that is difficult to write about, as so much that one can say risks misinterpretation: issues that are only touched upon might cause experts to see the treatment as trivial; topics analyzed with the lens of science might cause some to perceive the treatment as lacking compassion for those who have suffered; and a focus on the biological causes of evil might be misperceived as an argument for the inevitability of our brutish nature, and the denial of our peaceful and cooperative side. I hope that these misconceptions will be avoided by my treatment of the topic.

I am also greatly humbled by the massive literature on evil, developed by scholars in theology, philosophy, and the social sciences. My decision to write a short book was not motivated by a desire to reduce this richness to a few footnotes, nor to push to the side the humanistic issues linked to the topic of evil. Rather, my goal was to use the wealth of insights from the past to inform a highly focused look at one aspect of evil, one that can, I believe, be enriched by the many new ideas and discoveries of the biological sciences. My desire to write a short and accessible book for a wide audience also means that I have left out exhaustive references, in-depth descriptions of our atrocities, and comprehensive engagement with the many theories that offer to explain evil; these can be found in the general references I cite in the footnotes at the end of each chapter, and in many other books and papers.

If this book has any impact on current discourse it is because it offers a minimalist explanation of evil, of how it evolved, develops within individuals, is nurtured by different cultures, and affects the lives of millions of innocent victims. I believe, as do many scientists, that deep understanding of exceptionally difficult phenomena requires staking out a piece of theoretical real estate with only a few properties, putting to the side many interesting, but potentially distracting details. This book is my attempt to extract and explain the core of evil, the part that generates the kind of variation that our history has catalogued and that our future holds. It is a description of evil, not a prescription for what we, both as individuals and as societies, should do.

Sincerely,

acknowledgements

I wrote this book while my cat, Humphrey Bogart, sat on my desk, staring at the computer monitor. Though he purred a lot, and was good value when I needed a break, he didn't provide a single insight. Nor did our other pets: a dog, rabbit, and two other cats. For insights, critical comments on my writing, comfort, and endless love and inspiration, there is only one mammal, deliciously wonderful, and without an evil bone in her body — my wife, Lilan.

Marc Aidinoff ... a Harvard undergraduate who joined me early on in this journey, digging up references, collecting data, arguing interpretations, sharing my enthusiasm, while offering his own.

Kim Beeman and Fritz Tsao ... my two oldest and closest friends. They have some of the richest minds around. Their knowledge of film, literature, and the arts is unsurpassed. Their capacity to bring these riches to the sciences, and share them with me, is a gift.

Noam Chomsky... for inspiration, fearless attacks on power mongering, and friendship.

Marc D. Hauser

Errol Morris... for heated discussion, camaraderie, and insights into evil through his cinematographic lens and critical mind.

Many colleagues, students, and friends provided invaluable feedback on various parts of the book, or its entirety: Kim Beeman, Kent Berridge, George Cadwalader, Donal Cahill, Noam Chomsky, Jim Churchill, Randy Cohen, Kevin Doughten, Jonathan Figdor, Nick Haslam, Omar Sultan Haque, Lilan Hauser, Bryce Huebner, Ann Jon, Gordon Kraft-Todd, Sarah Lippincott, Errol Morris, Philip Pettit, Lisa Pytka, Michael Schneider, Richard Sosis, Fritz Tsao, Jack Van Honk, Wendy Wolf, and Richard Wrangham. A special thanks to Dan Dennett, David Livingstone Smith, Robert Trivers, and Steven Pinker for commenting on the book at a time when I needed honest, critical advice.

For moving the book through Amazon, I would like to thank Dan Slater.

Lastly, many thanks to my agent and friend John Brockman for standing by me during challenging times.

table of contents

prologue: the problem of evil

In reality, there is no such thing as "eradicating" evil tendencies.... [T]he deepest essence of human nature consists of instinctual impulses which are of an elementary nature,... and which aim at the satisfaction of certain primal needs.

— Sigmund Freud

It is a fact that humans destroy the lives of other humans — strangers, friends, lovers, and kin — and have been doing so for a long time. These cases are unsurprising and easily explained: we harm others when it benefits us directly, fighting to win resources or wipe out the competition. In this sense we are no different from any other social animal. The mystery is why seemingly normal people torture, mutilate, and kill others for the fun of it — or for no apparent benefit at all. Why did we, alone among the social animals, develop an appetite for gratuitous cruelty? This is the core problem of evil. It is a problem that has engaged scholars for centuries and is the central topic of this book.

Evildoers have many personalities. Some are cruel for cruelty's sake. Some believe that extreme violence is the only way to secure resources or defend sacred values. Some inspire others to do their dirty work. And some stand by and watch as others carry out horrific acts of violence, unwilling —though not unable — to intervene. You might think that these different behaviors require different explanations. I suggest that they all stem from a single psychological recipe that is part of every human mind but of no other mind in the animal kingdom. This is a stripped-down account of evil, one that explains how it grows within some individuals and how it uniquely evolved in our species.

The idea I develop is that evildoers are made in much the same way that addicts are made. Both processes start with unsatisfied desires. Whether it is a taste for violence or a taste for alcohol, drugs, food, or gambling, individuals develop cravings but find the desired experience less and less rewarding — a separation between desire and reward that leads to excess. To justify the excess, the psychology of desire recruits the psychology of denial, enabling individuals to immerse themselves in a new reality that feels right. Whereas addicts cause great harm to themselves by indulging in excessive consumption or expenditures, evildoers cause great harm to others by indulging in excessive or gratuitous cruelty. Whereas addicts deny their drug dependency or their obesity, evildoers deny the moral worth of their victims or invent a reality that presents them as dangerous threats. The cruelty carries no moral weight because the victims have been dehumanized or conceived as dangerous. The combination of unsatisfied desire and denial is a recipe for evil. Like the addict's search for ever more satisfying means of consuming or spending, evildoers search for ever more satisfying and creative ways of harming others.

This perspective, I suggest, explains not just the pathology of the sadist or the sexual predator but the actions of "ordinary" individuals who perpetrate unimaginable cruelties. It also illuminates the evolution of our capacity for evil, which, I will argue, evolved as an incidental consequence of our brain's unique design. This is an idea developed in somewhat similar ways by the philosopher David Livingstone Smith in his book *Less than Human*, and by the social psychologist Roy Baumeister in his book *Evil.* Unlike the brains of other animals, where circuitry specialized for one function slavishly serves that function, our brain circuitry works in harmony to serve a variety of novel functions. Thus, when we dehumanize other human beings — thinking of them, say, as vermin or parasites — and then torture them without guilt, we have connected brain areas involved in recognizing objects, determining moral standards, and justifying actions with brain areas involved in emotion, reward, motivation, and aggression. This is just one of many ways we can combine and recombine thoughts and emotions to create new ways of seeing the world. The point here is that this mental flexibility did not evolve to serve evil; rather, evil was enabled as an incidental consequence of our brain's unique design.[1]

Once the capacity for evil was in place in the human brain, it could be harnessed to serve a useful and adaptive function. By carrying out costly, over-the-top acts of violence, individuals signaled their ability to waste resources simply because they could — because they had the power or the wealth to do so. These displays sent credible messages of ongoing and impending terror to victims, freezing them in their own fear. This explanation for costly signaling, proposed by the evolutionary biologist Amotz Zahavi and developed in interesting ways by others, is one way of interpreting the paradoxical, gob-smacking episodes of gratuitous cruelty carried

---

[1] The difference between human and animal brains, and especially the distinction between dedicated modules serving one function and interconnected modules working to serve multiple functions, has been highlighted by other authors, including especially the philosopher Daniel Dennett, in *Consciousness Explained* (New York: Little, Brown, 1991), and the archaeologist Steven Mithen, in *The Prehistory of the Mind* (London, U.K.: Thames & Hudson, 1996). These and other authors emphasize that language was essential in forging the connection between modules. But, as I will discuss in chapter 3, language itself is based on interconnected modules, including those dedicated to phonology, semantics, and syntax. It is thus more likely that the connections were in place before language, providing benefits in thinking that went far beyond the parochial style of other animals.

out by otherwise civilized people.[2] It is an idea I develop further in chapter 3.

If my explanation for how evil develops in individuals and how it evolved is correct, it suggests that each of us has the potential to engage in cruel acts against innocent others. Equipped with the gift of imagination, we all entertain goals that are out of reach either because of personal limitations or because of constraints imposed by our own and others' moral standards. Tempted to achieve such goals, we may morally disengage, sometimes consciously and sometimes unconsciously. When we morally disengage, we enable a process of false justification for our actions, including self-deception and the dehumanization of others. On this view, everyone is capable of engaging in gratuitous cruelty because the ingredients that make up the recipe for evil are part of human nature, part of our uniquely evolved brains.

To understand evil is neither to justify nor excuse it, reflexively converting inhumane acts into mere accidents of biology or the unfortunate consequences of bad environments. To understand evil is to clarify its causes. In some cases, understanding entails recognizing that

---

[2] See especially Amotz Zahavi and Avishag Zahavi, *The Handicap Principle* (New York: Oxford University Press, 1997). On conspicuous consumption, see Thorstein Veblen, *The Theory of the Leisure Class* (London, U.K.: Macmillan, 1899). On credible threats, see Thomas C. Schelling, *The Strategy of Conflict* (Cambridge, MA: Harvard University Press, 1960).

a perpetrator suffers from brain damage or a developmental disorder and thus lacks self-control or awareness of others' pain — mitigating factors that influence legal decisions and should influence public perception as well. In other cases, it entails recognizing that a perpetrator was sound of mind yet knowingly caused harm to innocent others and relished the act. By describing and understanding an individual's character with the tools of science, we are more likely to make appropriate assignments of responsibility, blame, punishment, and future risk to society.

Homo sapiens, the knowing and wise animal, has logged an uncontested record of atrocities, despite moral norms prohibiting such actions. No other species has abducted children into rogue armies and then killed those who refused to kill, tossed infants into the air as targets for shooting practice, gang-raped women and forced them to carry the enemy's fetus to term, and burned people to death because more humane forms of killing were deemed less politically effective (or less enjoyable). These horrific acts have been universally described as "evil" by scholars, clerics, journalists, filmmakers, and novelists, as well as by those who have survived the mayhem. Despite the pervasiveness of such atrocities, some thinkers see them as rare defects of human nature, unfortunate malignancies that have metastasized within our species' essential goodness.[3] Yet others, such as the philosopher Luke Russell, recognize their ubiquity but see no value in labeling them as evil, because they merely represent actions along a continuum of moral wrongs.[4] It may well be a pointless debate. Even if one is unimpressed by the enormous number of people — approximately 80 million — who were tortured, brutalized, maimed, and senselessly killed on the watch of the most egregious dictators of the past hundred years — Idi Amin, Francisco Franco, Adolf Hitler, Kim Jong-il, Slobodan Milošević, Pol Pot, Josef Stalin, Charles Taylor, and Mao Zedong — this brief sampling of history points to a common pattern, cutting across continents, cultures, and economies. It is a pattern that warrants serious attention and explanation and that I will address in this book.

There is merit to using the word "evil" to describe certain human acts as long as one is clear about the concept and its defining features. This is the path I have chosen, building on the recent philosophical writings of

---

[3] See, for example, the philosopher Philippa Foot's *Natural Goodness* (New York: Oxford University Press, 2001) and the literary critic Terry Eagleton's *On Evil* (New Haven, CT: Yale University Press, 2010).

[4] See Luke Russell's essay "Is evil action qualitatively distinct from ordinary wrongdoing?" *Australian Journal of Philosophy* 85(4): 659-77 (2007). For a counterargument in favor of the distinctiveness of evil, see Todd Calder "Is evil just very wrong?" *Philosophical Studies* 163(1): 177-196 (2013).

Colin McGinn, John Kekes, and David Livingstone Smith and the work of psychologists Roy Baumeister and Ervin Taub. As I see it, *evil arises when innocent victims are subjected to gratuitous cruelty by individuals who either directly intend such excessive harm or allow it to happen when they could have prevented it.* This view of evil includes specific means (gratuitous cruelty), consequences (excessive harm), causes (intentions, desires, and goals), and potential benefits, both short- and long-term. These features require explanation, sketched below and developed in the chapters that follow.

Our commonsense understanding of evil, as portrayed in movies or novels as well as in media reports, typically centers on individuals who use gratuitous violence against others, often an excessive number of others. In the majority of genocides, for example, the perpetrators rarely seem satisfied with persecuting a subset of an ethnic group — or with just killing their victims. They slash or burn them to death, making sure that the "cleansing" leaves no survivors and no traces of the doomed minority. Eliminating an alien creed, say, or reducing competition for scarce resources by painlessly killing a segment of the hated population is not an option. Unlike sadists or serial killers who, because of their diseased minds can't stop themselves, mentally sound perpetrators of genocide or other atrocities could stop but choose not to. The question I will try to answer is: What leads seemingly sane people to use — or allow others to use — such horrific means and target such massive numbers, when simpler means and smaller numbers would suffice?

I will also attempt to explain why some individuals derive pleasure from seeing others destroyed, whereas others destroy lives without feeling any emotion at all.[5] When individuals use excessive methods of harming others, sometimes it is a "fix" for a diseased mind that has developed a predilection for inflicting pain. Sometimes it is the result of a long process of de-sensitization, in which cruelty yields no personal reward at all because hurting others has become routine. And sometimes it is the result of a long-term political strategy in which the use of extravagant and seemingly unnecessary means of hurting others is designed to stifle opposition, leaving surviving witnesses in a state of terror, uncertain as to where such cruelty might stop. I will discuss each of these situations, highlighting genes that shift individuals' perception of risk, self-control, and anticipated pleasure, as well as environments that encourage selfishness and lack of empathy for those who are different.

Students of human nature often end up in futile disagreements about the causes of behavior because they confuse explanations of how things work with explanations of

---

[5] For an explicit philosophical argument for the connection between pleasure and evil, see Colin McGinn's *Ethics, Evil, and Fiction* (New York: Oxford University Press, 1997).

how they evolved. These disagreements arise in scholarly discussions of language, of singing, of sex, of violence, even of eating. If you explain the human sweet tooth by describing the response of the brain's reward areas to sugar, this is not an alternative explanation for the one that identifies the selective pressures that favored fruit consumption among our primate ancestors. As originally explained by the Nobel laureate and ethologist Niko Tinbergen, explaining any aspect of behavior requires understanding not just how it works but how it evolved. Both explanations are critical to my account of evil. Part I of this book focuses on how individuals develop the capacity for evil. Part II explains how our species, uniquely, may have evolved this capacity. Here I provide a brief overview of the book's main ideas and the evidence I will present to support them.

Individuals develop into evildoers when unsatisfied desires accumulate and combine with a denial of reality, causing them to see others as morally worthless or dangerous. This proposal breaks down the capacity for evil into its minimal form, a recipe born out of only two psychological ingredients: desire and denial. When these combine, there is little to prevent people from engaging in gratuitous cruelty and much to inspire it.

Every one of us has desires to acquire resources and to seek out experiences because of the pleasure they bring. Our desires motivate us, sometimes to fulfill our own needs, sometimes to help others. We all desire good health, satisfying relationships, and knowledge of the world around us. Some also desire great wealth and power. Every culture has its signature vision of what counts, including money, land, livestock, wives, and subordinates. The desire system motivates action in order to achieve rewarding experiences. Some of our actions have benign, even beneficial consequences for ourselves or the welfare of others; some have malignant consequences for others and costly consequences for ourselves. Some desires can be satisfied; some cannot. To understand both outcomes, we need to understand how desire works.

Scientific studies pioneered in the 1950s by James Olds and Peter Milner and developed over the past two decades by Morten Kringelbach and Kent Berridge reveal that pleasure, and our desire to obtain it, consists of three different processes: *wanting, liking,* and *learning.*[6] We, together with the rest of the animal kingdom, typically want things we like and like things we want. For example, many monkeys and apes rely on a diet consisting primarily of fruit. These species are therefore motivated to find fruit as it is critical to their survival. These primates want fruit. These primates also like fruit, as evidenced

[6] See especially Olds, J. & Milner, P. (1954). Positive reinforcement produced by electrical stimulation of septal area and other regions of rat brain. *Journal of Comparative Physiology and Psychology* 47: 419-27; Berridge, K. (2009). Wanting and liking: Observations from the neuroscience and psychology laboratory. *Inquiry 52*(4): 378-98; Kringelbach, M. L. & Berridge, K. C. (2009). Towards a functional neuroanatomy of pleasure and happiness, *Trends in Cognitive Science* 13(1): 479-87.

by their emotional responses, including high-pitched coos and chirps when they see it. These primates don't have to learn to like fruit, but they do have to learn which objects are fruits and among these, which are edible and which are toxic. This coupling between wanting, liking, and learning is obviously adaptive. It would be odd if we or other animals consistently wanted things we didn't like or liked things we didn't want. But because these are distinct systems, they can be decoupled, either experimentally in the lab or naturally as a result of addiction. Addicts want more and more but like less and less the experience of getting what they want. When an individual develops an addiction — whether to alcohol, drugs, food, gambling, sex, shopping, or smoking — the wanting system ends up decoupled from the liking system, leaving the desire unsatisfied. The result is a stockpile of unsatisfied desires and an increasing resort to excess. Faced with this situation, individuals often turn to denial, persuading themselves that they are svelte, non-dependent or in need of another closetful of shoes. The idea I develop in Part I is that we initiate the recipe for evil when, as with other addictions, our unsatisfied desires for power — based in resources or ideological domination — accumulate to create cravings for more. When denial accompanies desire, progress in completing this recipe accelerates.

Every one of us engages in denial, negating certain aspects of reality in order to manage painful experiences, project a more powerful image, or justify a particular action. But like desire, denial has consequences both

beneficial and costly for self and others, a point powerfully developed by the evolutionary biologist Robert
Trivers in his book *The Folly of Fools*. When we listen to
the news and hear of human-rights violations, we often
shut off our emotions to create an illusion of peace.
When surgeons slice into human flesh, they turn off their
natural human empathy, treating their patient's body as
a mechanical device. Before we confront a challenging opponent in an athletic competition or in war, we
pump ourselves up, tricking ourselves into believing that
we're tougher than we are and our opponents weaker.
Denial enables us to motivate and justify actions in
cases where we are conflicted. So doctors who deny
the moral worth of others can be persuaded to carry
out heinous operations for "the good of science" or the
preservation of "racial purity." Military commanders in
denial of an opponent's strength can lead their troops
to annihilation. Soldiers who deny their enemies' moral
worth, thinking of them as less than human, can hack
into their flesh, not only deaf to the screaming but exhila-
rated by it. Political leaders can deceive their followers
and themselves into seeing the members of another
ethnic or cultural group as dangerous to the society's
sacred values, thereby justifying a purge. Individuals
in denial can reject various other aspects of reality to
enable acts of enormous cruelty.

Seen in this way, our capacity for evil is potentially
as great as our capacity for love and compassion. Evil
is part of human nature, a point noted long ago by the
philosopher Immanuel Kant in his treatise *Religion Within*

the Boundaries of Mere Reason. The idea that evil is part of human nature, or "innate" as Kant described it, doesn't mean that there is a gene for evil or an evil organ. It also doesn't mean that each of us will exercise the capacity for evil, combining unsatisfied desires with denial to cause excessive harm. Individual differences, arising both from our biology and our experiences, will make some of us more likely to do so. This leads us to the second *how* question: How did the human capacity for evil evolve?

The capacity for evil originally evolved as an incidental consequence of our unique intelligence, but once in place provided significant benefits to those who expressed it as a display of power. Like the brains of other animals, our brains consist of circuits that evolved to perform specific functions, such as recognizing objects, quantifying the value of an experience, and feeling pleasure or pain. Unlike those of other animals, our brains freely combine these single-purpose circuits to create multipurpose systems. For example, we combine the systems for object recognition, value, and emotion, allowing us to perceive other humans as objects (and thus tradable as commodities or dispensable as waste) while feeling good about making others feel bad. But to say that our capacity for evil originally evolved as a by-product of the brain's combining power is not to say that the capacity for evil lacks utility for human survival. On the contrary, and as I develop in chapter 3, once the capacity for evil evolved, it enabled individuals to display costly and gratuitous acts of violence as a means of intentionally intimidating others. These intentional acts are different from the uncontrollable, involuntary acts arising from disordered minds, though the consequences are the same — the torturous destruction of others.

These ideas about the evolution of evil distinguish questions of origin from questions of current function. They are questions I will address using different kinds of evidence. To assess origins, and in particular the claim that human brains are unique in some way, we look to comparative studies of other animals. To assess current function, and in particular the idea that cruelty results in evolutionary payoffs, we look to comparative studies of human cultures and the survival consequences of intimidating others.

All animals show highly specialized abilities to solve problems linked to survival. Honeybees perform dances to tell others about the precise location of nutritious nectar, providing an information highway that lowers the costs of individual foraging challenges. Meerkats teach their young how to hunt dangerous but energy-rich scorpion prey, providing an education that bypasses the risks of trial-and-error learning. Each of these extraordinary behaviors is used for solving one and only one problem — except in humans. Animal thoughts and emotions are single-purposed, focused on a single functional problem, deploying what the biologists Dorothy Cheney and Robert Seyfarth call "laser beam intelligence."[7] Human

---

[7] See for instance their books *How Monkeys See the World* (Chicago: University of Chicago Press, 1990) and *Baboon Metaphysics* (Chicago: University of Chicago Press, 2007).

thoughts and emotions are multi-purposed, even though many originally evolved to solve one specific problem. Thus, humans unconsciously wrinkle their noses and pull back their lips into an expression of disgust that communicates information about disease-ridden and toxic substances, thereby lowering the costs of sickness to others who might be exposed. But this same expression also appears when we witness morally abhorrent actions, such as sodomy, bestiality, or various forms of torture. Thus, an emotional expression that originally evolved to signal a physically toxic and dangerous object is connected to our moral sense, in order to signal toxic and dangerous individuals or even abstract concepts such as racism or religious intolerance. Combining thoughts and emotions is what enables a brain capable of evil. It enables individuals with unsatisfied desires to deny aspects of reality in the service of destroying innocent lives.

What the sciences reveal is that our brain's capacity for combination was realized by evolutionary changes in the number of wired-up brain areas. As these connections increased, our brains were able to move beyond the narrow and specialized functions of particular areas and thereby solve a broader range of problems. Though we don't know precisely when these changes occurred, we do know they occurred after our split from the other great apes — orangutans, gorillas, bonobos, and chimpanzees.

See also Marc Hauser (2009). The possibility of impossible cultures. *Nature* 460: 191-96.

We know this from looking at the brains of these species and also from examining how they use tools, communicate, cooperate with, and attack one another. Not only are there fewer connections between regions of the brain in these primates, their thinking in various contexts is single-purposed, faithfully dedicated to the solving of one specific problem.

Empowered by our new, massively connected brain, we alone migrated into and inhabited virtually every known environment on Earth and some beyond, inventing abstract mathematical concepts, conceiving grammatically structured languages, and creating glorious civilizations rich in rituals, laws, and philosophies. Our massively connected brain also equipped us for evil, but only as an incidental consequence of other adaptive capacities, including the ability to harm others for the purpose of surviving and reproducing.

Many social animals fight over resources using ritualized behaviors to assess their opponents and minimize the personal costs of injury. Most of this fighting involves nonlethal aggression, with losers intimidated or injured but able to walk away. There are, however, three situations in which individuals deploy lethal aggression, and in each case, as noted by the psychologist Victor Nell and the anthropologist Richard Wrangham, there are significant asymmetries in power and significant survival benefits.[8] First, predators kill prey, enabled by substantial asymmetries in their weaponry, designed for killing (think viper venom, cheetah canines and hawk talons). Second, in a wide variety of species, adults kill vulnerable infants. Adult males kill infants sired by other males in the group, a strategy designed to wipe out the competition and bring adult females back to a cycle in which they can conceive again. Adult females kill infants when times are tough and parenting is too costly. Third, in only four animal groups — ants, wolves,

---

[8] Nell, V. (2006). Cruelty's rewards: The gratifications of per-petrators and spectators. *Behavioral & Brain Sciences 29*(3): 211-24; Wrangham, R. W. & Glowacki, L. (2012). Intergroup aggression in chimpanzees and war in nomadic hunter-gath-erers: Evaluating the chimpanzee model. *Human Nature* 23(1): 5-29; Wrangham, R.W. & Peterson, D. (1996). *Demonic Males: Apes and the Origins of Human Violence* (Boston MA: Houghton Mifflin); Wrangham, R.W. (2010). Chimpanzee violence is a serious topic: A response to Sussman and Marshack's critique of Demonic Males: Apes and the origins of human violence. *Global Nonkilling Working Papers* 1:29-50.

For conflicting views on the relevance of this work to under-standing the origins of human warfare, see Ferguson, R.B. (2011). Born to live: Challenging killer myths. In: *Origins of Altruism and Cooperation*, Developments in Primatology: Progress and Prospects, 36(3): 249-70; Hart, D. & Sussman, R.F. (2009), *Man the Hunted: Primates, Predators, and Human Evolution* (New York: Basic Books); Horgan, J. (2102) *The End of War* (San Francisco CA: McSweeney's).

lions, and chimpanzees — adults regularly form coalitions to kill adults from a neighboring competitor if, and only if, the attacking coalition outnumbers the competitor by a considerable margin. The apparent rarity of lethal aggression in the animal kingdom, and especially its restricted context, is indicative of single-minded thinking. This pattern stands in striking contrast to the pattern of killing seen in human societies.

Like other animals, humans kill when there are imbalances of power, including asymmetries in weaponry, physical size, and numbers. As Wrangham has noted, we see such imbalances in youth gangs who, like chimpanzees, form stealth coalitions to sneak up and attack lone victims.[9] Unlike other animals, however, our own species kills in fundamentally different ways. We kill not only on massive scales but also in a wide variety of contexts, attacking strangers, friends, lovers, kin, the old and the young — and not just when there is an imbalance of power but also when power is matched. We kill with and without immediate provocation, for personal gain or no reason at all. We have invented methods to kill painlessly and to inflict prolonged, excruciating pain. Only *we* are known to be repeatedly cruel for cruelty's sake. These differences suggest that something changed over the course of evolution, both in the ability to carry out lethal aggression and the advantages of doing so. What changed, I suggest, is our brain's ability to combine

---

[9] Wrangham, R.W. (1999) "Evolution of coalitionary killing," *Yearbook of Physical Anthropology* 42: 1-30.

thoughts and emotions to create new ways of solving old problems. This change enabled us to overwhelm our competitors by impressing them with excessive violence. As Zahavi explains in his theory of costly signaling, only those individuals who can afford to waste resources will display with such exuberance. By extension, individuals with enough wealth and power can readily afford to exhibit gratuitous acts of cruelty. These costly signals empower the perpetrator by striking fear and trembling in the victims. Seen in this way, gratuitous cruelty is strategic, and *anything but* the outcome of a disordered mind. It is this kind of healthy evildoer who represents the central puzzle I explore in this book.

In Part I, we will focus on the recipe for evil, examining both the psychology of desire (chapter 1) and the psychology of denial (chapter 2). In Part II, we will look at the evolutionary history of evil. Chapter 3 examines what makes us unique relative to other animals and assesses the possible adaptive function of seemingly wasteful acts of extreme violence. Chapter 4 explains why some people are more likely to develop into evildoers than others, based on a combination of nature and nurture.

My goal is to guide you on a journey into evil. It is a journey that describes our evolutionary past, our present state of affairs, and the prospects for our future. It is as much a story about you and me, here and now, as it is about our long-ago ancestors and our descendants to come. It is an account of the nature of moral decay and the prospects for moral growth. It is not an account that will teach us how to banish evil from the world; I don't believe that's possible. Rather, it is an account that will help us understand why some individuals acquire an addiction to feeling good by making others feel bad, often while destroying blameless lives under the banner of virtue. It may well prompt us to recognize our own vulnerabilities and monitor the variety of ways in which desire and denial combine to create an evilicious mindset.

Recommended reading

• There are numerous books about evil, most written by philosophers, theologians, historians, political scientists, and legal scholars. The following are books about evil written by scientists. I have learned a great deal from them, and many of their ideas powerfully enrich the pages between these covers:

Baumeister, R. F. (1997). *Evil: Inside Human Violence and Cruelty.* New York: W. H. Freeman.

Baron-Cohen, S. (2011). *The Science of Evil: On Empathy and the Origins of Cruelty.* New York: Basic Books.

Oakley, B. (2007). *Evil Genes: Why Rome Fell, Hitler Rose, Enron Failed and My Sister Stole My Mother's Boyfriend.* Amherst, NY: Prometheus Books.

Staub, E. (2010). *Overcoming Evil: Genocide, Violent Conflict, and Terrorism.* New York: Oxford University Press.

Stone, M. H. (2009). *The Anatomy of Evil.* Amherst, NY: Prometheus Books.

Zimbardo, P. (2007). *The Lucifer Effect: Understanding How Good People Turn Evil.* New York: Random House.

• For a comprehensive discussion of evil by philosophers and historians, with an eye to the relevant science, see:

Kekes, J. (2007). *The Roots of Evil.* Ithaca NY: Cornell University Press.

Livingstone Smith, D. (2011) *Less Than Human: Why We Demean, Enslave, and Exterminate Others.* New York: St. Martin's Press.

McGinn, C. (1999). *Ethics, Evil and Fiction.* Oxford: Oxford University Press.

Shermer, M. (2004). *The Science of Good and Evil.* New York: Henry Holt Inc.

• On killing throughout history, see:

Grossman, Lt. Col. D. (1995). *On Killing: The Psychological Costs of Learning to Kill in War and Society.* New York: Little, Brown.

Wrangham, R.W. & Peterson, D. (1996). *Demonic Males: Apes and the Origins of Human Violence.* Boston, MA: Houghton Mifflin.

part i:
one recipe

one: runaway desire

The desire of being believed, or the desire of persuading, of leading and directing other people, seems to be one of the strongest of all of our desires.

                              – Adam Smith

In 2003, Marwan Abu Ubeida (his pseudonym) started preparing for what he believed would be the happiest day of his life. He couldn't imagine wanting or desiring anything else. His preparations involved intense physical and psychological training, including frequent prayers. When Marwan prayed, he asked for a blessing of his mission, purification of his soul, and the ability to reunite with his brothers in the afterlife. He also read historical accounts of the great martyrs of the past. These readings, he believed, would give him the strength to patiently wait for the day when he could act and satisfy his ultimate desire to die.

Marwan was a Sunni Muslim jihadi in waiting, waiting for the green light to transform himself into a suicide bomber. All facets of his desire to die were clear. As captured by *Time* magazine correspondent Bobby Ghosh[10], Marwan wanted to achieve his mission with perfection: "If I am lucky, my body will be vaporized. There won't be anything left of me to bury." He wanted to meet other suicide bombers in his afterlife: "We made a pact that we would meet in heaven." He only wanted to harm Iraqi infidels and American soldiers: "I pray no innocent people are killed in my mission." And he had no interest in being recognized by his fellow Iraqis: "The only person who matters is Allah."

Marwan is one of thousands of martyrs who, over the course of history, have sacrificed their lives in the name of a great cause. Their desires seem, in one way, pure: to support a set of sacred beliefs and values. And yet, the consequence of their selfless action is often excessive harm to innocent others. Because most of us can't imagine decorating our bodies with dynamite, and then pressing the self-destruct button, we tend to think that suicide bombers are either uneducated, mentally deranged, or religious fanatics. Though some suicide bombers carry such resumes, many, including past and present members of al-Qaeda, Hezbollah, and Hamas are college educated, mentally healthy,

---

[10] For accounts of suicide martyrs, see: Ghosh, B. (2005) Inside the mind of a suicide bomber. *Time Magazine*, June 26; Ginges, J., Atran, S., Sachdeva, S. & Medin, D. (2011) Psychology out of the laboratory: the challenge of violent extremism. *American Psychologist 66*(6), 507–519.

and have no religious education. As summarized by the anthropologist Scott Atran and his colleagues, suicide martyrs often come to religious fanaticism late in life, but such fanaticism is not what pushed them into a life of extreme violence. Instead, the best predictors of who will turn to this kind of terrorism as a way of life are social networks of family and friends, concern over national humiliation, and the perception of inappropriate interventions by foreigners. What is important here is the powerful role that sacred values have on some individuals, and how such values can drive a deep desire to harm others, even if this means harming oneself. For many, this is an example of desire run amuck. For others, it is an example of self-sacrifice for the good of the group. Whichever way one leans, suicide bombing is a clear case where an individual's desire leads to horrific consequences for innocent others.

Desire is the first ingredient in the recipe for evil. In this chapter I lay the foundation for understanding how desire works, explaining how it arises as a thought or feeling, how it motivates action, and how it spins out of control leading to addictions. I will explain how we acquire and nurture the desire to obtain valuable resources such as food, water, money, and mates, but along the way often harm other individuals or allow harm to happen. The puzzle that I will explain is how a benign process that shapes our aesthetic preferences and motivates our capacity to acquire beautiful paintings, delicious food, and attractive partners can turn into a process that motivates us to kill, using excessive means to reach excessive ends, sometimes for personal pleasure and sometimes with no feeling at all.

The desire for pleasure

Imagine that scientists have just announced the discovery of a center in the brain that manages our experience of pleasure. Imagine further that they have invented a consumer device called *Bliss* that, for only $49.99, enables you to ramp up or down the activity in this pleasure center. Want more out of your dinner, movie, tennis stroke, work, or sex? Flip the *Bliss* switch. Want to buffer yourself from the pain of ostracism, a romantic breakup, or a colonoscopy? Flip the *Bliss* switch. Would you buy *Bliss*? Before you answer this question, think about potential side effects. Did you think about the possibility that you might become addicted to *Bliss* or worse, either destroy the feeling of pleasure altogether or end up in a never-ending quest for satisfaction, each dollop of pleasure leaving you wanting a bigger dollop next time around. This may seem like science fiction, but it's closer to non-fiction.

Over fifty years ago, scientists implanted electrodes into a region of the rat's brain called the *nucleus accumbens*. The electrodes were connected to a switch. If the rat pressed the switch, the electrode turned on and so too did the nucleus accumbens. The rats indeed pressed, over and over again, some at a rate of two thousand presses per hour, with no external reward or threat of punishment. Pressing the switch was the reward, or at least the vehicle to a rewarding experience. Pressing the switch was addictive.

Soon after this discovery, clinicians started using the same brain stimulation technique to treat individuals with neurological complications, including Parkinson's patients suffering from loss of motor control, patients experiencing sustained pain, Tourette's patients suffering from motor tics and obsessive-compulsive problems, and even a patient in a coma who had lost, but then slowly recovered the capacity to name and grasp objects. As in the rat work, a clinician implanted an electrical pulse generator within a targeted brain region. When the generator turned on, it stimulated activity in previously malfunctioning regions. But sometimes it stimulated much more than the doctor planned.

Two patients suffering from chronic pain developed profound *addictions* to the stimulation. In addition to relatively successful pain reduction, both patients also experienced an enhanced desire for sex, including erotic feelings. One of these patients self-stimulated so often that she forgot to wash, change clothes, and adhere to family commitments. Though it is unclear why damping down pain resulted in ramping up the desire for sex — as opposed to other rewarding experiences such as eating food — it is clear that certain brain areas control how much we want particular rewarding experiences. This work also suggests that specific brain regions figure prominently in the pathway to addiction, uncontrollable cravings that are toxic to self and others. This work suggests that brain areas associated with desire can run out of control. We need to understand this process as it is a critical step in the path to causing gratuitous cruelty.

To understand how the brain motivates us to *want* some things but not others, how it creates the experience of *liking*, and how it enables us to want things we like by *learning* about the world, we turn to experiments on nonhuman animals, brain scans of healthy humans, the mechanics of mind-altering drugs, conscious and unconscious influences on our choices, and the forces that lead individuals to develop uncontrollable urges to eat, drink, snort, shoot up, and gamble[11]. This is the

---

[11] The evidence behind the elements of desire: Berridge, K. C. (2009). Wanting and liking: Observations from the neuroscience and psychology laboratory. *Inquiry, 52*(4), 378-398; Kringelbach, M., & Berridge, K. C. (2009). Towards a functional neuroanatomy of pleasure and happiness. *Trends in Cognitive Sciences, 13*, 479-487; Kringelbach, M., & Berridge, K. C. (2010). The functional neuroanatomy of pleasure and happiness. *Discovery Medicine, 9*(49), 579-587; Olds, J. (1956) Pleasure centers in the brain. *Scientific American*, 195:105–16; Olds J & Milner P. (1954) Positive reinforcement produced by electrical stimulation of the septal area and other regions of rat brain. *Journal of Comparative Physiology and Psychology* 47:419–27; Portenoy, R. K., Jarden, J. O., Sidtis, J. J., Lipton, R. B., Foley, K. M., & Rottenberg, D. A. (1986). Compulsive thalamic self-stimulation: a case with metabolic, electrophysiologic and behavioral correlates. [Case Report]. *Pain, 27*(3), 277-290.

evidence that scientists have gathered to explain the nature of wanting, liking, and learning.

In both humans and other animals, we can understand how the wanting system works by measuring what individuals approach when given a choice, as well as how much effort they are willing to exert while approaching and gaining access to a particular object or experience. For example, in studies that explore whether captive animals have sufficient housing conditions, an experimenter presents a choice of rooms, one consisting of the typical housing environment and the others with additional goods believed to be of interest. To enter a given room requires opening a door. To determine how much an individual really wants what is in another room, the experimenter makes it extremely difficult to open each door by adding a tight spring. In studies of captive hens and mongoose, individuals exerted considerable effort to open some doors but not others: hens rammed into doors opening onto a chipped wood floor, whereas mongoose did the same for a swimming pool of water. These are items they want, but are deprived of in captivity. Humans deprived of basic rights — food, water, living space — exert similar efforts to obtain these goods. Effort exerted is a measure of wanting, whether in hens, mongoose, or humans.

What about liking? It may seem that there are no clear objective ways to measure liking because it is a subjective experience. My likes are my own. You can't possibly know how intensely I like French cheese or reading George Eliot novels or teaching children who want to learn. And if you can't know what it is like for me to like things, then we can't possibly know what it is like for a hen or mongoose to like things. There are, however, ways of measuring liking and disliking that are reliable, objective, and consistent across species. In many animals, including human babies who can't speak and human adults who have lost this capacity due to brain injury, there are distinctive behaviors that are consistently linked to positive experiences and others linked to negative ones. For example, in mice, monkeys, and human babies, tasting something sweet like sugar causes a lot of lip licking, whereas tasting something bitter such as quinine causes mouth gaping, nose twitching, and arm flailing. These similarities suggest that evolution has been conservative, preserving the same means of expressing likes and dislikes in different species. These similarities have enabled scientists to understand how the brain systems involved in wanting and liking can change together or separately, even though they can't help us understand the harder problem of what, subjectively, it is like for a given mouse, monkey or man to like something.

Understanding which components of a system are necessary or sufficient for it to operate requires isolating the components, turning some on and others off. Take the batteries out of a portable radio and it doesn't work. Are batteries necessary? Yes, unless one has an electrical cord to plug in the wall. Are batteries sufficient? No, because without a transistor, antenna, and speaker, the batteries have nothing to drive. The same logic applies when we look at brain systems and specifically, the engine behind wanting and desiring. If, following damage or a lesion to a specific brain area, an individual no longer executes a specific behavior, say eating and enjoying the food consumed, it is reasonable to infer that this area is necessary for eating and enjoying food consumption; it may or may not be sufficient. The same logic applies to molecular manipulations in which scientists silence or turn on specific genes with known functions.

The cognitive neuroscientists Susana Peciña and Kent Berridge took a population of mice and silenced a gene that controls the neurochemical dopamine, causing the circulating levels in the brain to increase[12]. The reason they targeted dopamine is because it plays a central role in regulating both the anticipation and experience of reward in invertebrates and vertebrates, revealing the signature of an evolutionarily ancient system. Compared with normal mice, these dopamine-plus mice consumed twice as much food and water, and learned much faster where food was located within a maze. But when it came to measuring licking as liking, the dopamine-plus mice were no different from normal mice. Dopamine is,

---

[12] Genetically engineered mice: Peciña S., Cagniard B., Berridge K.C., Aldridge J.W., Zhuang X. (2003). Hyperdopaminergic mutant mice have higher "wanting" but not "liking" for sweet rewards. *Journal of Neuroscience* 23:9395–402; Peciña, S., Smith, K. S., & Berridge, K. C. (2006). Hedonic hot spots in the brain. *The Neuroscientist 12*(6), 500–511.

therefore, essential for the wanting system but not the liking system. This conclusion has been supported by many other studies, of mice and men, in the context of eating and drug addiction — two topics that I will discuss momentarily.

To understand what rodents like, Peciña and Berridge injected an opioid drug — similar to opium from poppy plants — into two brain regions associated with reward — the nucleus accumbens and the ventral pallidum. More precisely, they injected the drug into areas called *hedonic hotspots* — zones tuned like pitchforks to particular kinds of stimulation, designed to jazz up liking. Following injection, individuals licked four times more often in response to sugar as the non-injected individuals, but showed no difference in wanting. They also showed less of a gag response to the nastiness of bitter quinine. Turning on these hedonic hotspots ramped up the pleasure of sweets and diminished the displeasure from bitters. Together, the Peciña and Berridge studies highlight the independence of wanting and liking, and the ways in which the brain — or a clever experimenter playing with it — regulates them.

How does the brain figure out what's hot and what's not, delicious or disgusting? It's one thing to desire a particular experience and another to derive a rewarding or pleasurable experience. But the world is not set up with labels that indicate which objects and events are delicious and which disgusting. Labeling is an active process, carried out by the organism, and constrained by initial biases that are part of every sensory ability, as well as experiences accumulated over a lifetime. All organisms start off life with biases that cause them to hear, smell, see, and taste some things better than others. These biases propel them toward some objects and away from others. This is why no human baby has to be taught to dislike bitter things and like sweet things. From the very first encounter, sugary solutions trigger tongue protrusions and licking, whereas bitter solutions trigger a gaping mouth. We have evolved, as have other animals, sensory systems that are tuned to prefer some things and dislike others right from the start. These initial biases guide learning, facilitating acquisition of new knowledge in some cases and making it almost impossible in others. This is where scientific evidence gains considerable interest, helping us understand how we develop anticipatory pleasures and past-oriented regrets, struggle to change from habitual rewards, and acquire irrational desires for experiences we no longer enjoy. It is the power and pull of these desires that lead us to seek their satisfaction.

Humans go to restaurants and bees to flowers because both are associated with food. Within these broad categories, there are good restaurants and flowers, as well as bad ones, where good and bad are determined by experience. The experience can be direct, as when food is actually consumed, or indirect, as when humans listen to an animated friend describe a restaurant's menu and bees watch a hive mate dance a description of the flower's location and quality. Once the association between food and location is established, simply

Marc D. Hauser

seeing the restaurant or flower triggers a cascade of neural and chemical activity in the brain linked to reward, including its anticipation. The restaurant and flower are cues that predict food. If you walked into your favorite restaurant and found that they sold fertilizer rather than food, you would be heartbroken. If you haven't been to the restaurant in a long time, but memorialized your previous experience as a gastronomic high point, you will be deeply disappointed if your first bite doesn't live up to the standards you anticipated. This mismatch between anticipated and experienced reward leads to a cascade of brain activity — indicative of an error. The primary engine driving the experience of reward, including predicting the timing and intensity of its occurrence, is the dopaminergic system, a network of brain areas that releases dopamine in virtually all animals, including humans[13].

---

[13] Dopamine in and out of control: Chen, T., Blum, K., Mathews, D., & Fisher, L. (2005). Are dopaminergic genes involved in a predisposition to pathological aggression? *Medical Hypotheses, 65*, 703-707; Di Chiara, G., & Bassareo, V. (2007). Reward system and addiction: what dopamine does and doesn't do. *Current Opinion in Pharmacology, 7*, 69–76; Doya, K. (2008). Modulators of decision making. *Nature Neuroscience, 11*(4), 410–416; Dreher, J.-C., Kohn, P., Kolachana, B., Weinberger, D. R., & Berman, K. F. (2009). Variation in dopamine genes influences responsivity of the human reward system. *Proceedings of the National Academy of Sciences (USA), 106*(2), 617–622; Everitt, B. J., Belin, D., Economidou, D., Pelloux, Y., Dalley,

There are several genes that control the expression of dopamine, a point I alluded to earlier when describing the work of Peciña and Berridge on mice. For each gene, there are variants that cause individual differences

J. W., & Robbins, T. W. (2008). Review: Neural mechanisms underlying the vulnerability to develop compulsive drug-seeking habits and addiction. *Philosophical Transactions of the Royal Society B: Biological Sciences, 363*(1507), 3125–3135; Everitt, B. J., & Robbins, T. W. (2005). Neural systems of reinforcement for drug addiction: from actions to habits to compulsion. *Nature Neuroscience, 8*(11), 1481–1489; Grigorenko, E. L., De Young, C. G., Eastman, M., Getchell, M., Haeffel, G. J., Klinteberg, B. A., Koposov, R. A., et al. (2010). Aggressive behavior, related conduct problems, and variation in genes affecting dopamine turnover. *Aggressive Behavior, 36*(3), 158–176; Johnson, P. M., & Kenny, P. J. (2031). Dopamine D2 receptors in addiction-like reward dysfunction and compulsive eating in obese rats. *Nature 13*(5), 635–641; Sabbatini da Silva Lobo, D., Vallada, H., & Knight, J. (2007). Dopamine genes and pathological gambling in discordant sib-pairs. *Journal of Gambling Studies, 23,* 421–433; Sharot, T., Shiner, T., Brown, A. C., Fan, J., & Dolan, R. J. (2009). Dopamine enhances expectation of pleasure in humans. *Current Biology, 19*(24), 2077–2080; Volkow, N. D., Wang, G.-J., & Baler, R. D. (2011). Reward, dopamine and the control of food intake: implications for obesity. *Trends in Cognitive Sciences, 15*(1), 37–46; Walter, N., Markett, S., & Montag, C. (2010). A genetic contribution to cooperation: Dopamine-relevant genes are associated with social facilitation. *Social Neuroscience* 6: 289-301.

in how much dopamine is expressed. Those variants leading to higher concentrations of dopamine are associated with greater risk-taking, as well as higher levels of compulsive gambling, eating, and substance abuse. This suggests that changes in the level of dopamine are associated with both changes in the anticipation of reward, and a diminished capacity for self-control. When we anticipate a highly rewarding experience and lack self-control, excessive consumption is a likely outcome. To determine whether the link between dopamine, reward anticipation, self-control, and excess is correct, we turn to experimental evidence.

To measure anticipated rewards, the cognitive neuroscientist Tali Sharot asked individuals to rate how happy they would be if they vacationed in eighty possible destinations. Although you may think that a Caribbean island retreat is ideal, whereas someone else imagines the cafes and museums of Paris, the interesting patterns arise when comparing each individual's choices — their ranking of the Caribbean relative to Paris and several other locations. Following this first round of ranking, some subjects took a placebo while others took L-dopa, a drug that selectively increases the release of dopamine. Later, these same subjects imagined what it would be like to actually vacation in these eighty destinations, and rated their imagined experience. Consistently, those on L-dopa felt they would be much happier. This suggests that dopamine isn't just associated with changes in our anticipation of and desire for a reward. Dopamine drives our experience of an anticipated reward, whether vacation, food, or drugs. This regulation of anticipated rewards is an adaptive process, as it enables us to set up expectations and determine when they are violated.

Changes in dopamine can also cause an uncalibrated experience of anticipated reward, an out-of- control desire for more. This occurs because surges in dopamine also impact the brain circuitry involved in self-control. When dopamine levels surge, either naturally or due to experimental intervention, the capacity for self-control loses its grip, leading to the kinds of excess seen in addictions to food, drugs, and gambling. This is a maladaptive process. Regulation of dopamine thus determines, in part, whether our desires are calibrated or uncalibrated, and thus healthy or unhealthy. Clinical populations help us understand this process.

Parkinson's patients suffer from brain dysfunction in areas that regulate the level of dopamine. In general, they have lower levels of dopamine and as a result show deficits in learning. These deficits are in large part due to their failure to connect actions with anticipated reward. Medically boosting dopamine in Parkinson's patients can improve learning, and as in Tarot's studies noted above, this occurs because they are more sensitive to the relationship between actions and rewards. But, in a significant number of Parkinson's patients, boosting dopamine with L-dopa results in pathological gambling, a classic problem of self-control. Those patients who are most vulnerable are those with a particular variant of a gene known as DRD4 that results in elevated levels of dopamine. In healthy populations, those with this same gene variant are also more likely to exhibit pathological gambling, as well as attention disorder deficit — a global clinical problem involving poor self-control that afflicts approximately 1 out of 20 children. These studies show that addictive behaviors are commonly associated with abnormally high levels of dopamine. Addictive behaviors reveal the relationship between poor self-control and excess.

Consider obesity, a global and rapidly increasing health problem today. To appreciate the numbers, let's focus on the United States, the country with the highest proportion of obese people in the world. Based on 2011 analyses, approximately 35% of adults and 20% of children under the age of 19 years were considered obese. These percentages convert to approximately 60 million obese people in the United States alone. Estimates suggest that by 2030, the percentage of obese people will rise to 42%, adding on about 32 million more individuals, and an additional financial burden of $550 billion due to health care costs alone. As health economist Eric Finkelstein notes, these costs are joined by others, including the billions of dollars lost due to lack of productivity, and the billions of dollars added for transportation costs, including lowered fuel efficiency and shifts in the size and seating dimensions of vehicles[14].

---

[14] Finkelstein, E.A., Khavjou, O.A., Thompson, H., Trogdon, J.G., Pan, L., Sherry, B., & Dietz, W. (2011). Obesity and severe obesity forecasts through 2030. *American Journal of Preventive Medicine*, 42(6), 563-570.

There are many pathways to obesity, just as there are for other addictive behaviors. Several researchers suggest, however, that obesity is caused by changes in the reward system and a diminished capacity for self-control. When obese individuals view images of food during a brain scanning session, they show lower levels of activity in the *striatum* than individuals of normal weight; the striatum is an area rich in dopamine and an essential part of the reward system. This pattern may, at first, seem paradoxical: how could those who eat to excess show lower rather than higher levels of activity in the striatum, and thus, an over-the-top experience of reward upon seeing food? The answer lies in the brain's response to excess consumption: overeat and the reward knob shuts down. Under normal conditions, this is an adaptive response as it prevents overdoing it. When this same system operates in the mind of an overweight person, it is a disaster. Though the feedback from the reward center is closed, the food addict still wants a fix. With motivation on high, he continues to seek food, hoping to satisfy his desire. This cycle continues, often with devastating consequences. It is a cycle that is shared with other addictions, revealing that the systems that break down when we turn to excess work similarly whether the target is food, drugs, alcohol, gambling, or, I suggest, violence. It is a malfunction that reveals what happens when you unhinge desire from the experience of reward.

The work on addictive behavior tells us that independently of how people get started on the path to fulfilling their desires, and whatever leads them to over-consume, consumption loses its luster. The brain is smart: excess is bad and so the reward system shuts off. This creates a major problem because the wanting system is still turned on, looking for pleasure in all the wrong places. The result: unfulfilled cravings.

The work on addiction provides a template for thinking about how individuals step onto a path that leads to excessive harms. In the same way that excessive eating gets going and going out of control when the dopamine system drives an irrational desire to want more and more food that is liked less and less, so too I suggest can excessive harm arise when wanting and liking part company. Individuals start with a desire to acquire wealth, to physically harm those who are unlike them, or taste the sweetness of revenge against someone who acted unfairly. These desires are often linked to a rewarding experience or the anticipation of one, a point I will soon support with evidence. But as such actions and their consequences accumulate, the pleasure derived diminishes. Liking is no longer part of the equation; but wanting is. The result: unsatisfied cravings blind individuals to the harms caused.

To develop the analogy between food, drug, alcohol and gambling addictions on the one hand, and addictions to violence on the other, I need to fill in several missing pieces, starting first with a richer description of the psychology of desire. Everything I have discussed in this section has focused on individuals and their core corporal needs for survival — or in the case of drugs and gambling, recreation. I haven't said a word about how desire works in the social arena, whether the same systems are in play when we compare our own desires and resources with others, or with other opportunities. When desire is motivated by what others have or have achieved, are the same processes in play as when we eat, drink, or gamble? These are important questions as the desire to accumulate great wealth or to harm others is often motivated by comparison shopping, assessing what others have relative to our own status. The most primal starting point for comparison shopping is the world of hierarchies, a world where the desire to dominate rules.

Power hungry

In social insects, fish, lizards, birds, rodents, whales, apes, or humans, males are bigger and bolder, more boisterous, brash and brazen, and more motivated to get into a brawl than females. Though biologists don't define the sexes based on these differences, they use them to understand what drives competition for valuable resources and what determines the criteria for dominance status. Biologists define the sexes based on differences in the gonads, the reproductive organs that generate eggs and sperm, and the corresponding effects of sex-specific selection on the mind, body, and behavior. Females are those with larger, more costly gonads, where cost is defined on the basis of how much energy is invested in production. Think eggs versus sperm. This difference sets up an immediate competition, especially for species that have parental care. Once you invest in a big expensive egg, you don't want to lose your investment. You want to protect it, avoiding harm and minimizing risk. On the other hand, if your investment is small, as is the case for sperm, you are not only freer to take risks, but favored by selection to do so.

These ideas about sexual selection started with Charles Darwin. One hundred years later, they were developed in detail by the evolutionary biologist Robert Trivers. Combined, they provide an explanation for why, in most species including our own, males compete with each other for access to females — the most valuable and limited resource — and why females are picky, expressing an aesthetic preference for males of a particular quality. Selection favors parts of the body and brain associated with dominant males and picky females. Dominant males win fights against other males, and thus gain greater access to females. Dominant males take risks and are more aggressive. Picky females hold out for the best males, those who provide the most desirable resources. Picky females are patient, waiting for males with good genes, access to prime real estate, and the protective skills and motivation to defend them and their young. These are qualities linked to high status. These are qualities associated with the ability to obtain and control resources. These are the qualities that females desire and are motivated to obtain.

What these basic evolutionary processes lead to is a tango of desires, a topic that will occupy us in chapter 3. For now, the key point is that across all socially living animals, study after study supports the conclusion that male desire for dominance swings with female desire for dominant males. This dance leads males to evolve ever more powerful means of winning, and pushes females to be ever more discriminating and demanding in terms of a male's qualifications as a mate[15].

---

[15] How dominance works: Boksem, M. A. S., Kostermans, E., Milivojevic, B., & De Cremer, D. (2012). Social status determines how we monitor and evaluate our performance. *Social Cognitive and Affective Neuroscience* 3: 304-313; Chiao, J. Y. (2010). Neural basis of social status hierarchy across species. *Current Opinion in Neurobiology, 20*(6), 803–809; Chiao, J. Y., Mathur, V. A., Harada, T., & Lipke, T. (2009). Neural basis of preference for human social hierarchy versus egalitarianism. *Annals of the New York Academy of Sciences, 1167*(1), 174–181; Deaner, R. O., Khera, A. V., & Platt, Michael L. (2005). Monkeys pay per view: adaptive valuation of social images by rhesus macaques *Current Biology, 15*(6), 543–548; Klein, J. T., Deaner, R. O., & Platt, Michael L. (2008). Neural correlates of social target value in macaque parietal cortex *Current Biology, 18*(6), 419–424; Liew, S., Ma, Y., & Han, S. (2011). Who's afraid of the boss: Cultural differences in social hierarchies modulate self-face recognition in Chinese and Americans. *PLoS ONE, 6*(2), 1–8; Watson, K. K., & Platt, M L. (2008). Neuroethology of reward and decision making. *Philosophical Transactions of the Royal Society B: Biological Sciences, 363*(1511), 3825–3835.

Given the push for increasing means of winning and discriminating, there is a premium on social information. Information about a competitor's weakness or the quality of a mate, provides an upper hand in fighting and reproducing. The neuroscientist Michael Platt carried out a clever series of experiments with male rhesus monkeys to explore how much individuals would pay to gain access to socially relevant information. To set the stage, keep in mind that rhesus monkeys are a hyper-aggressive species with a strict dominance hierarchy: high ranking individuals outcompete lower ranking individuals for food, places to rest, and mates. If someone tries to step out of line, and gets caught, the costs can be extreme. In the heat of the mating season on the island of Cayo Santiago, Puerto Rico, I have witnessed high ranking male rhesus monkeys pin down lower ranking males trying to sneak a mating, rip into their groin with knife-like canines, and extract one testicle. Hyper-aggressive, competitive, and costly for the loser.

In Platt's experiment, each monkey watched a slide show with viewing options akin to pay-per-view television. On a given trial, they could watch one of two images for as long as they liked, each viewing option associated with a particular amount of juice. For each pair of images, one delivered more juice than the other. Given that these monkeys were thirsty, they should prefer more juice over less juice. If monkeys have no interest in the images per se — because they have no value — then their viewing preferences should be strictly determined by where they can get the most juice. If, on the other hand, the images have value, and some images are more valuable than others, then they may be willing to look at an image that delivers less juice over an image that delivers more. This is paying for viewing, and a proxy for their motivation and desire for juice. Evidence that rhesus monkeys value the social information that comes from the image over the juice itself would be a surprising result given that juice is a primary reward whereas the image is only a secondary reward — an indication or cue of things to come.

Consistently, these male monkeys had two favorite channels, preferring those showing pictures of high ranking individuals and close-ups of female derrieres. They preferred these over images of low ranking individuals, despite the fact that their preferred choice often cost them the opportunity to drink more juice.

Platt's findings show that rhesus monkeys are motivated to acquire information about socially relevant situations, including information about dominance and sex. Their motivation and desire to obtain this information is high, as evidenced by the fact that they are willing to pay a cost — similar in important ways to the hens bashing open heavy doors to get chipped wood. Keeping an eye on a dominant is of value as dominants pose a threat, especially one staring at you. Keeping an eye on a female's rear end is also of value as it can signal sexual receptivity: in rhesus monkeys, as in many other monkeys and apes, the area around the vagina either swells, turns red, or both when females are ovulating.

This is important information for males, guiding their desire to court and mate with willing females.

Humans also value social information, devoting many hours in a day to gossiping and people-watching, whether in magazines, tabloids, reality TV, or real life. This information showcases what we have relative to others, a difference that will be exaggerated in societies with dominance hierarchies.

People living in hierarchical societies should relish information showing that a competitor has lost resources, whereas those living in egalitarian societies should be motivated to redress the imbalance. To test this possibility, the social psychologist Joan Chiao used survey information to establish two groups of individuals, one with preferences for hierarchical societies and the other with preferences for egalitarian societies. Chiao placed individuals from these two groups into a brain scanner and showed them pictures of people experiencing pain — an experience that often triggers empathy. Two areas, both associated with the personal experience of pain and the perception of pain in others, were highly active; this is the brain's way of representing empathy. But these areas were less active in those who preferred hierarchies than those who preferred egalitarianism. This finding, as Chiao notes, is consistent with the idea that in an egalitarian society, empathy for others' well-being is essential. In egalitarian societies, seeing someone who has less or is being harmed by another, should motivate a desire to redress the imbalance and reduce the harm. In a dog-eat-dog hierarchical society, where dominants outcompete subordinates and inequities are part of life, concern for those at the bottom is a sign of weakness.

Chiao interpreted her results as evidence that cultural influences can shape brain activity, leading some to develop a desire for dominance and inequities, whereas others develop a desire for equality. It is possible, however, that these individuals started out life with structural differences in brain anatomy and function, and these differences led some to prefer societies that champion inequalities while others prefer those that support equality. Chiao's work can't distinguish between these alternatives. Nonetheless, her studies nicely show that patterns of brain activation can heighten our sensitivity to what others have, what we desire, and how some of our desires can flexibly change in response to what others have. And if the addictive model I have been pushing is more broadly applicable to all of our desires, then our desire for status can turn into an unhinged drive for power that is difficult to satisfy.

I'll have what she's having

One of the most famous lines in movie history was delivered by Estelle Reiner in *When Harry Met Sally,* a comedy produced by her son Rob Reiner. While Estelle is seated at a table in a delicatessen, Sally — played by Meg Ryan — fakes having an orgasm to show Harry — played by Billy Crystal — that he can't tell the difference between fakes and the real deal. Overhearing Ryan's performance, Estelle turns to the waiter and says "I'll have what she's having." This is comparison shopping, cashing in on someone else's subjective experience to guide our chosen experiences.

Orgasms and eating are two of the most pleasurable experiences in life, whether you live in Tokyo, Toronto, Toulouse, Tehran or Timbuktu? I doubt any healthy human adult would debate this. What can be debated is what counts as the ultimate orgasm or food experience. It can be debated both among friends and inside our own minds, influenced by personal experience and our knowledge of what else is available, or might be. It is this comparative perspective that drives our desire for more, with the consequence, intended or incidental, that others have less. To nail this point, let me first step back to a particularly insightful study of food before stepping forward to studies of comparative shopping in the social market. Though food and social status are different commodities, the underlying psychology of comparison is remarkably similar.

Consider potato chips. As a snack, potato chips generate a revenue in the United States of about $6-7 billion dollars each year, relying on the slicing and frying of about 2 billion pounds of potatoes. These facts make clear that most Americans love potato chips, and are motivated to consume them. The social psychologist Carey Morewedge and his collaborators ran an experiment to find out how much people love potato chips, and whether their anticipated fondness for this delicious crisp changes in the face of other options[16]. Subjects sat at a table in front of a bowl of potato chips and an alternative food that was visible, but out of reach. The alternative was either a highly undesirable snack such as sardines, or a highly desirable one such as Godiva chocolate. After subjects contemplated what it would be like to eat each of these foods, they then rated how much they would enjoy them. This approach is like Sharot's study that I described earlier where subjects rated how much they would enjoy different vacation destinations, but without the comparison between one clearly good and one clearly bad spot. Both studies focus on the anticipation of a pleasurable experience, as opposed to the actual experience of it.

Subjects' ratings of potato chip pleasure soared when sardines were in view and plummeted when compared with chocolate. Context matters. What we unambiguously anticipate as a delicious experience when there is nothing else on the table, loses or gains in anticipated pleasure when the table fills up with other delectable or disgusting alternatives.

What's happening to our pleasure detector in the potato chip experiment and especially our anticipation of reward? Are we incapable of understanding what makes us happy, unable to figure out what is or is not delicious?

---

[16] Potato chip tastes: Morewedge, C. K., Gilbert, D. T., Myrseth, K. O. R., Kassam, K. S., & Wilson, T. D. (2010). Consuming experience: Why affective forecasters overestimate comparative value. *Journal of Experimental Social Psychology*, *46*(6), 986–992.

Or are we fickle? What Morewedge's experiment suggests is that deliciousness, like ugliness, stubbornness, and obsequiousness, is a judgment, judgments are always relative or comparative, and as such, based on some standard that is either present in the moment, stored away in our memories, or anticipated in the future. When Estelle Reiner uttered her famous line, she was using Meg Ryan's orgasmic expression of delight as a comparative metric. When we compare food, wine, pretty faces, sporty cars or land that our neighbor controls, we recruit our brain's resources, especially the circuitry involved in attention, emotion and memory. Whether we say that potato chips are the best snack, or better than sardines, we have made a comparison that requires our attention, our capacity to keep at least two items in memory, and a way of emotionally tagging each of the items — one yum, one yuck. This comparison-shopping taxes our mental resources, recruiting them away from the job of evaluating one snack, and leading to a distorted evaluation of desirability or worth.

Morewedge's experiments point to a mismatch between how delicious something is and how delicious we think it will be, or how delicious we thought it was. It reveals a distortion in our capacity to anticipate — or *forecast* in the words of social psychologist Daniel Gilbert — how we will feel, and in particular, how much we will like the experience. This represents a kink in the psychology of desire: we expect the system that links wanting and liking to be well honed so that we end up really wanting things we really like. This distortion arises not only in the context of eating, but in a variety of other contexts including the social world.

Consider revenge. When someone transgresses over the border of social norms, either harming us or those we care about, we often seek revenge, motivated to even things up. We often imagine that revenge will make us feel better, providing a honey hit to the brain that will satisfy our desire to redress an imbalance. In brain terms, this honey hit would be expressed as dopamine given the essential role that this chemical plays in the anticipation of reward. But do we consistently experience this feeling of reward when we follow through on a plot of revenge or, as Sir Frances Bacon noted over three hundred years ago, might "A man that studieth revenge, keeps his own wounds green, which otherwise would heal, and do well."[17] Might our desire for revenge inoculate us against healing, creating an illusion that we will feel better? If so, revenge is like addiction, with wanting unhinged from liking.

To understand how revenge works, the social psychologist Kevin Carlsmith set up an experimental game, allowing each subject within a group to contribute money to a public good.[18] At the end of one

---

[17] Spedding, J., Ellis, R. L., & Heath, D. D. (Eds.). (1858). *The Works of Francis Bacon* (Vol. 6). London: Longman; cited in Carlsmith et al. (2008), p.g. 1324.

[18] The paradox of revenge: Carlsmith, K. M., Wilson, T. D., & Gilbert, D. T. (2008). The paradoxical consequences of revenge. *Journal of Personality and Social Psychology, 95*(6), 1316–1324.

round of contributions, the bank (an experimenter distributing the money) multiplied the total contribution by a pre-determined amount, divided this total by the number of players, and then redistributed this amount to each player. In this game, if each player contributes to the common pot, everyone comes out with the highest returns. However, the best strategy for an individual is to cheat, holding on to the initial endowment while reaping the rewards of everyone else's generous contributions. Those who opt out of cooperation in a public good situation stand to benefit, especially if there are no punitive consequences. In Carlsmith's set up, some could pay to punish, some witnessed the consequence of another's punitive act, and some played in a game with no punishment.

When given the opportunity to punish the cheater, most people punished. Everyone, both punishers and non-punishers alike, expected punishment to feel good. They were wrong. Both punishers and those who witnessed punishment felt worse, with the act of punishment compounding the bad feelings. The fact that the witnesses felt worse, as opposed to better, may seem at odds with our experience of schadenfreude, of enjoying another's misery. Shouldn't the witnesses have rejoiced upon discovering that the offenders were slapped with a punitive fine? In our own personal experience with schadenfreude, as well as in studies that I will soon discuss, people rejoice over someone else's misfortune but this news has no direct bearing on them. In Carlsmith's experiments, the witnesses learn of a misfortune, but the offender's defection directly harms the witness in terms of money lost. These results suggest that although punishment may feel good, this benefit may not make up for the lost income.

Everyone in Carlsmith's experiments also believed that punishment would cause people to think less about the offender, flushing the bad thoughts from their memories. They were wrong again. Punishers, but not those who simply witnessed punishment, ruminated more about the selfish offenders. Rumination led to more bad feelings. These bad feelings led to more rumination, giving birth to a vicious cycle of feeling bad and ruminating about those who cheated them of some money. Rumination also affected the level of punishment: those who ruminated more, and felt worse about it, also punished more heavily. Rumination heightened the comparative difference in resources. Rumination focused attention on the wrongdoer, akin to a police investigation with a spotlight beamed on to the criminal's face. Rumination caused the ruminator to think more about what he didn't have, and how he was cheated. Rumination maintained the bad feelings in memory.

Carlsmith's findings are paradoxical and disturbing. Paradoxically, they suggest that in some situations, our expectations about the feeling of punishing an immoral act are inverted from the feelings we feel following punishment: rather than feeling a happy high, we feel a depressing low, often accompanied by increasing anger. In the context of punishing a free-rider who stiffed the

group, everyone expects to feel good, but many end up feeling angry instead. The entire polarity of the emotion has switched, with rumination and anger dominating our thoughts. Like Morewedge's study of food, our anticipation of pleasure is also distorted for revenge. This is a dangerous state to enter. Faced with the strong belief and desire that revenge should feel good, but lacking any confirmation, we are moved to find new evidence and as shown in other experiments by Carlsmith, to dole out more punishment to satisfy the feeling of just deserts, the feeling that the punishment fits the crime. With anger at the helm, there is one highly probable outcome: the delivery of increasingly severe levels of punishment, driven by the desire to feel good about retribution. This is precisely the pattern I described above for obesity: the wanting system continues to search for liking and reward, but fails and thus continues its search. Whether it is an unsatisfied desire for food or revenge, the unfortunate consequence is an escalation to excess, fueled by a failure to obtain what we want.

The great leveler

We are often envious of those who have what we desire, whether it is good looks, money, a warm supportive family, or a better tennis stroke. Envy, an emotion that is anchored in comparison shopping, motivates us to change our looks, find careers that will improve our finances, seek relationships that will provide additional support, and pick up a few extra tennis lessons to win the next match. Unfortunately, envy can quickly turn malignant, as desire and a deep sense of inferiority about our personal deficiencies transform into insatiable cravings to acquire whatever is necessary to gain superiority. Envy thus wears two masks, at times motivating self-improvement, at others destructive toward those who have more. Envy attempts to level the playing field, taking from the Haves to deliver unto the Have-Nots.

Envy emerges out of our sense of fairness, fueled by competition. It is part and parcel of a hierarchical society. When we envy someone, we have detected a difference or inequity between our own condition and that of another. We want what someone else has, presumably because we like or think we like what they have. Wanting and liking are in harmony. Recognition of the inequity fuels competition to redress the imbalance. This sense of fairness appears early in child development and changes in systematic ways as a function of a culture's norms[19].

---

[19] Envy and fairness: Almas, I., Cappelen, A. W., Sorensen, E. O., & Tungodden, B. (2010). Fairness and the development of inequality acceptance. *Science, 328*(5982), 1176–1178; Blake, P. R., & Mcauliffe, K. (2011). "I had so much it didn't seem fair": Eight-year-olds reject two forms of inequity. *Cognition, 120*(2), 215–224; Dvash, J, Gilam, G, Ben-Ze'ev, A, Hendler, T, & Shamay-Tsoory, S.G. (2010). The envious brain: The neural basis of social comparison. *Human Brain Mapping* 31(11): 1741-1750; Fehr, E., Bernhard, H., & Rockenbach, B. (2008). Egalitarianism in young children. *Nature, 454*(7208),

Marc D. Hauser

The economist Ernst Fehr set up an experiment to explore how fairness develops in young children ages 3-8 years old. Fehr was especially interested in when children recognize a disparity or inequity in the distribution of resources and what they are willing to do, if anything, to redress the imbalance. Each child first learned that they would play against a partner of the same age who was either from the same school or a different school; Fehr set up the school distinction to look at the possibility, seen repeatedly in experimental studies and real world

1079–1083; van de Ven, N., Zeelenberg, M., & Pieters, R. (2009). Leveling up and down: the experiences of benign and malicious envy. *Emotion, 9*(3), 419-429; Henrich, J., Boyd, R., Bowles,S., Gintis,H., Fehr, E., Camerer, C., McElreath, R., Gurven, M., Hill, K., Barr, A., Ensminger, J., Tracer, D., Marlow, F., Patton, J., Alvard, M., Gil-White F., and N. Henrich (2005) 'Economic Man' in cross-cultural perspective: Ethnography and experiments from 15 small-scale societies. *Behavioral and Brain Sciences*, 28, 795-855; Henrich, J., McElreath, R., Barr, A., Ensminger, J., Barrett, C., Bolyantz, A., Cardenas, J. C., et al. (2006). Costly punishment across human societies. *Science, 312*, 1767–1770; Smith, R. H., & Kim, S. H. (2007). Comprehending envy. *Psychological Bulletin, 133*(1), 46–64; van de Ven, N., Zeelenberg, M., & Pieters, R. (2010). Warding off the evil eye: When the fear of being envied increases prosocial behavior. *Psychological Science,* 21(11): 1671-1677; Van Dijk, W.W. Ouwerkerk, J.W., Goslinga, S., Nieweg, M., & Gallucci, M. (2006). When people fall from grace: reconsidering the role of envy in Schadenfreude. *Emotion, 6*(1), 156-160.

situations with adults, of heightened cooperation among those from the same group and heightened hatred and aggression toward those outside. Though each child knew about his or her partner's age and school affiliation, they never saw their partner. Each child therefore knew only that they were playing with *someone* from their school or *someone* unfamiliar to them.

All children played three different games. In each game, one child decided how to distribute a fixed amount of candy to his or her partner. In the *prosocial* game, the decider took either one candy and gave one to the partner or took one candy and gave nothing to the partner. If children are sensitive to inequities and want to share in order to make things fair, they should pick the 1-1 option; picking the 1-0 option doesn't affect the decider, but dings the partner. In the *envy* game, the decider has a choice between 1-1 and 1-2. As in the prosocial game, the decider gets the same amount of candy with both options, but preserves equity with 1-1. Picking 1-1 also reveals that the child has an aversion to others having more, even when there is no personal cost. In the third, *sharing* game, the decider has a choice between 1-1 and 2-0. Here again, the decider gets candy in both cases, but the 2-0 option tempts the desire for more, both personally and relatively. On the one hand, a greedy child will want more candy, and so 2 wins over 1. But picking the 2-0 option also leads to a greater difference with the partner — critical for comparison shopping — while robbing them of an opportunity for any candy. If fairness prevails, deciders should pick 1-1.

If selfishness prevails, motivated by competition, they should pick 2-0.

Fehr uncovered two key results. Across all three games, younger children played more selfishly than older children, but independently of age and the game played, all children played more fairly with familiar than unfamiliar schoolmates. These results, together with several other studies, show that children are sensitive to the distribution of goods at an early age, but with important developmental changes in play. There is a tendency for children to both recognize inequities early in life, but to act selfishly when possible. The envy game provides a beautiful illustration. When another child could receive more, children, especially young ones, rejected this option even though it wouldn't cost them directly; the decider always gets just one candy in this game. Though no one has yet fully explained what causes this developmental shift from more to less selfishness, most agree that it is driven in part by maturation of brain regions guiding self-control, together with social factors that make young children increasingly aware of and sensitive to their own and others' reputations. Young children are more impulsive and selfish, and this leads to higher levels of inequity when they are asked to distribute valuable goods. Fehr's studies also show that playing fair is not just about the distribution of resources, but about who gets them. Early in life, children have already carved up the world into those they know and those they don't. This division drives their thinking and feeling, and in cases like this, their sense of fairness. It is a division that I will discuss in greater detail in chapter 2.

Fehr's research, and the majority of studies on the child's developing sense of fairness, focus on children living in large Western societies. Most of the work on fairness in adults is similarly focused on such societies. The precise structure of these societies may directly impact how individuals decide when to share, what commodities enter into the distribution, and whether sharing depends on effort invested, needs, or power. As noted in the last section, those who support an egalitarian society are more likely to feel empathy toward those in pain than those who support a hierarchical society. Individuals who are more empathic are also more altruistic. Hunter-gatherer societies tend to be more egalitarian, and highly cooperative. These differences predict further differences in how those living in small-scale societies, including the hunter-gatherers and subsistence farmers of Africa, Asia, and South America, should respond to unfair exchanges, and thus, whether they envy those who have more. If envy is lower in these societies, individuals should be less bothered when others get more, and thus be less interested in leveling the playing field.

The anthropologist Joseph Henrich and his colleagues presented a set of bargaining games to adults living in different small-scale societies across the globe. Though the subjects in this study played a number of different games, the basic goal was similar to those deployed by Fehr in his studies of children: under what conditions do individuals choose to share an unequal distribution of resources, and what are the consequences of selfish acts?

Consider the ultimatum game. One individual decides how to distribute a fixed amount of money to an anonymous partner. The partner has two options: keep what is on offer or reject it. Rejection costs both players as they leave empty handed. Rejection is both an expression of sour grapes for what could have been — a fair offer — and punishment for selfish behavior. In large scale societies, offers typically range from 40-50% of the initial pot, and rejections are common for offers less than about 20%.

Across the globe, most people in these small scale societies offered some amount of the initial pot. Across the globe, most people rejected really low offers. This suggests the universal signature of fairness and a desire to level those who try to get ahead. Cultures differed with respect to how much they shared and their threshold for rejecting offers. Some societies offered, on average, close to 40%, while others offered as little as 15%. Some societies accepted virtually all offers, whereas others rejected both low and even high offers. Even in more egalitarian societies, therefore, there is sensitivity to unequal distributions. Even in egalitarian societies, there is a willingness to punish those who act unfairly, those who take more than what others see as their fair share. Even in egalitarian societies there is a desire to prevent the Haves from having too much.

Let's take stock of the discussion thus far. Our sense of fairness is part of human nature, appearing early in development, but guided by experience toward a particular cultural form. The reason why I have discussed fairness is because it often plays an essential role in our experience of envy. Envy grows when we detect an uneven distribution of resources, wishing we were members of the haves as opposed to the have-nots. Envy grows when desire combines with competition, motivating a departure from the group of have-nots. Envy is yet another way in which we can accumulate unsatisfied desires, seeing the rest of the world as always having more. This perception of inequity can drive competition and hatred.

Studies of the brain show how envy is generated from the psychologies of desire and competition. When healthy subjects sit in a brain scanner and learn about other individuals who have what they desire, there is considerable activity in the *anterior cingulate*, and more activity in those who feel more envious. This is not the envy center of the brain. There is no such area. But the recruitment of the anterior cingulate in other social situations helps us understand what is going on more generally in the case of envy. There is significant activity in the anterior cingulate when we experience pain from social exclusion, but not when we witness such pain in others. It is one of the areas that was activated in Chiao's work on the differences in pain empathy between individuals who prefer egalitarian as opposed to hierarchical societies. The anterior cingulate is also involved when our minds are pulled in two different directions, a situation that arises when we are forced to choose between two conflicting moral options – for example, a duty to save the lives of many versus the prohibition of killing one person to save the lives of many. There is a common thread here that unites these different experiences. Like our experience of social exclusion, envy is also a form of social pain and to some, deeply painful, as it spotlights our deficiencies. Envy also represents a situation in which our positive sense of self conflicts with the negative sense of self engendered by social comparison. Our brain informs us that we are less accomplished when compared with others. Envy is socially imposed pain generated by comparison shopping.

The experience of envy highlights what we don't have, which fuels the system of desire, which seeks satisfaction. Unsatisfied, wanting keeps hunting for pleasure. The experience of schadenfreude delivers some prey — a morsel of joy that emerges from witnessing someone who is worse off.

O Schadenfreude

When the envied fall down or experts fail, we often perversely enjoy the knock out. This is schadenfreude, a German word that describes the joy we feel in witnessing another's misfortune. Though the emotion is universally understood, recognized in our written records at least as far back as Aristotle, the German language is one of only a handful of languages to capture the feeling in a single word, combining the word for harm (schaden) with the word for joy (freude). Like envy, schadenfreude is a social, comparative emotion. It erupts when those we envy fall down, when someone we dislike meets his comeuppance, and when a misfortune is deserved. And like envy, schadenfreude presents two faces, one elevating and virtuous, the other deflating and divisive. We should feel good when a person is caught crossing a moral line, and when justice is served. Such feelings not only reinforce our own adherence to moral norms, but encourage us to punish those who transgress. But these same feelings can emerge when we harm those who have been dehumanized. Dehumanizing another, as we shall see in chapter 2, provides one method of justifying morally abhorrent behavior. When members of another group are seen as parasites, eradicating them is not only justified but necessary for survival. Our desire for eradication continues until there is an extinction. Let's look more closely at how schadenfreude works, and the evidence to support these general observations[20].

---

[20] Under the hood of schadenfreude: Bushman, B. & Baumeister, R. (1998). Threatened egotism, narcissism, and direct and displaced aggression: does self-love or self-hate lead to violence? *Journal of Personality and Social Psychology* 75(1): 219-229; Baumeister, R. (2005). The lowdown on self-esteem. *Los Angeles Times*, http://articles.latimes.com/2005/jan/25/opinion/oe-baumeister25; Combs, D. J. Y., Powell, C. A. J., Schurtz, D. R., & Smith, R. H. (2009). Politics, schadenfreude, and ingroup identification: The sometimes happy thing about a poor economy and death. *Journal of Experimental Social Psychology*, 45(4), 635–646; Leach, C.W., & Spears, R. (2008).

Marc D. Hauser

Schadenfreude, like envy, can trigger self-evaluation, looking inside of ourselves to assess our net worth relative to others. We know from a large body of studies, several carried out by the social psychologist Roy Baumeister, that when an individual's sense of self-worth

---

A vengefulness of the impotent: the pain of in-group inferiority and schadenfreude toward successful out-groups. *Journal of Personality and Social Psychology, 95*(6), 1383-1396; Shamay-Tsoory, S.G., Fischer, M., Dvash, J., Harari, H., Perach-Bloom, N. & Levkovitz, Y. (2009). Intranasal administration of oxytocin increases envy and schadenfreude (gloating). *Biological Psychiatry, 66*(9), 864-870; Shamay-Tsoory, S.G, Tibi-Elhanany, Y, & Aharon-Peretz, J. (2007). The green-eyed monster and malicious joy: the neuroanatomical bases of envy and gloating (schadenfreude) *Brain, 130*, 1663-1678; Singer, T., & Steinbeis, N. (2009). Differential roles of fairness- and compassion-based motivations for cooperation, defection, and punishment. *Annals of the New York Academy of Sciences, 1167*, 41–50; Singer, T., Seymour, B., O'Doherty, J. P., Stephan, K. E., Dolan, R. J., & Frith, C. D. (2006). Empathic neural responses are modulated by the perceived fairness of others. *Nature, 439*(7075), 466–469; Takahashi, H, Kato, M, Matsuura, M, Mobbs, D, Suhara, T, & Okubo, Y. (2009). When your gain is my pain and your pain is my gain: Neural correlates of envy and schadenfreude. *Science, 323*, 937-939; Van Dijk, W. W., Ouwerkerk, J. W., Wesseling, Y. M., & van Koningsbruggen, G. M. (2011). Towards understanding pleasure at the misfortunes of others: the impact of self-evaluation threat on schadenfreude *Cognition and Emotion, 25*(2), 360–368.

is threatened, especially those individuals with more narcissistic and overly confident personalities, aggression often follows. The more personally threatened we feel by another individual or group, the more pleasure we should feel when they suffer. The psychologist Wilco van Dijk tested this idea with an experiment. Subjects filled out a questionnaire that they believed evaluated their intellectual strengths. Upon completing the questionnaire, some were told that they had utterly flopped, scoring in the lowest 10% of all subjects, while others were told that they performed brilliantly, scoring in the upper 10%. Next, all subjects read a scenario in which someone suffers a misfortune. For example, in one scenario, a student rents an expensive car to show off at a party, but then drives the car into a river, not only damaging the car but requiring the fire department to tow it out. Those whose sense of self-worth was threatened by the abominable test score were more likely to say that they felt good about the misfortune, including smirks and laughter in response to the show off who submerged his rented car, as well as other similar cases.

van Dijk's results show that schadenfreude serves the beneficial function of hoisting our own self-worth. This feeling even arises in cases where we have no connection with the injured party. When our self-worth has been challenged, for whatever reason, we feel better knowing that someone else is worse off, regardless of context or direct relevance. We evaluate our self-worth in much the same way that we evaluate potato chips: it's all relative to someone or something else.

Marc D. Hauser

I have explained that our desire to see others suffer can grow, especially when the desire is not satisfied by feeling good. Carlsmith's work on revenge and punishment, discussed earlier, provides an example: when people punish wrongdoers they expect to feel better, but ruminating on the experience causes them to feel worse and seek greater satisfaction. This cycle can drive a greater and greater desire to hurt others. Cycles such as these are often ignited and fueled by ideology, including especially differences in our perception of who is like us and who isn't. This distinction, as we learned earlier, guides children's sense of fairness, causing them to favor equity with in-group members and inequity among out-group members. Later in life, religious and political beliefs further fortify our favoritism toward in-group members, often leading to seemingly irrational desires to see those not like us suffer.

The psychologist Richard Smith explored whether an individual's political convictions influenced the intensity of schadenfreude when witnessing a member of another party suffer, including cases where society at large also suffers. Smith initiated the study prior to the US Presidential elections in 2004 involving Republican George W. Bush and Democrat John Kerry; during this period, the Republicans controlled both the House of Representatives and the Senate. Each subject — all college undergraduates — provided information about party affiliation and strength of support for the policies and beliefs of their party. Next, each subject read and provided reactions to short newspaper articles describing tragicomic moments for the two candidates, one in which Bush fell off a bicycle he was riding, and the other involving Kerry wearing a bizarre space outfit during a visit to NASA. Last, subjects read and reacted to an article describing job losses and the economic downturn facing the nation — an article meant to capture an objective cost to all members of society, irrespective of party affiliation.

Unsurprisingly, Democrats expressed more pleasure from reading about Bush's bicycle accident, whereas Republicans were more joyful over Kerry's bizarre space suit. Surprisingly, Democrats also expressed pleasure reading about the economic downturn and more pleasure than the Republicans who were more likely to express negative feelings about this situation. Thus, despite the fact that the economic downturn hurt everyone, the Democrats expressed pleasure over the added damage this inflicted on the Republicans — whom they held responsible — and conversely, the added benefit it brought to the Democrats who could wag their fingers. The Democrats' focus on the Republicans' fall utterly trumped the economic crash that impacted all US citizens, irrespective of party affiliation.

In a second study, Smith found that Democrats experienced more schadenfreude than Republicans over the number of casualties reported out of the Iraq war, even though Iraqis were certainly not preferentially targeting Republicans. The pleasure they experienced — seemingly irrational given the atrocities of war — was entirely driven by the fact that this was a war sponsored by a Republican government, and thus, the fatalities could be blamed on the Republicans. From a Democrat's perspective, even though everyone loses when soldiers die in war, it is a bigger loss for Republicans and thus, a bigger gain for Democrats. With schadenfreude, as with envy, it is all about comparison shopping. It is all about satisfying our desires to gain status relative to others, whether the other is a member of another political party, religious organization, or sports team. This can result in highly irrational responses, as desire for personal satisfaction catapults to the top, dominating all other considerations.

As I mentioned earlier, schadenfreude, like envy, is tied to notions of fairness: we feel good when someone who has more than us suffers a misfortune. To explore the relationship between fairness and schadenfreude more closely, the cognitive neuroscientist Tania Singer set up a study involving healthy men and women. Subjects first played a bargaining game for money against an unfamiliar partner; prior to the game, and unbeknownst to the subject, Singer set things up so that the partner played either fairly or unfairly. After the game, each subject entered a brain scanner, and watched their partner receive a painful shock to the hand.

Predictably, Singer discovered that both men and women liked the fair players better than the unfair players, and showed more empathy when the fair players were shocked. Proof of empathy was read off the images of brain activation, especially the brain circuitry known to be involved in pain empathy, and mentioned earlier in my discussion of Joan Chiao's work on social hierarchies: the insula and anterior cingulate. More surprisingly, Singer discovered that the level of activity in this area of the brain was reduced when men — but not women — saw unfair players shocked into pain; like Chiao's subjects, the competitive men showed little compassion for the pain experienced by a competitor. Singer also observed that in men — but not women — there was increased activity in the nucleus accumbens — an area mentioned earlier on that, in rats, monkeys and humans is consistently associated with the experience of reward and liking. The more individual men expressed a desire for revenge, the greater the activity in this reward area. Seeing a competitor go down is rewarding, and at least in these experiments, especially rewarding for men. As noted earlier, this difference fits well with the Darwinian expectation that males are the more competitive, status-oriented sex.

Singer's findings are joined by many others showing that the nucleus accumbens, together with other reward areas, are activated in a wide variety of situations in which we gain from others' pain. But because these same areas also respond to non-social, non-comparative experiences such as eating, we come back to a critical point in this chapter (and discussed in detail in chapter 3): areas that evolved for one function are readily recruited for others, especially in a highly connected brain like ours that readily combines different thoughts and emotions to create different ways of seeing the world. As long as something makes us feel good, whether it is winning, eating, social comparison, or harming another, the reward areas of the brain turn on.

Schadenfreude is one of the mind's ambassadors, enabling us to journey from a state of inferiority to superiority. It enables "imaginary revenge"[21] in the words of philosopher Friedrich Nietzsche. Like envy, it is highly adaptive, focusing our attention on inequities. Like envy, it also has maladaptive consequences, rewarding us when the inequity is not only addressed, but results in another's failure and misery. This sets up a potentially dangerous transformation from feeling good when we *witness* others feel bad to feeling good when we *make* others feel bad.

Up to now, I have largely focused on the feelings and thoughts associated with desire. Most of the desires I have discussed are fairly reasonable, even justifiable. If someone has hurt us personally, or hurt someone we care about, it is reasonable for us to want some kind of retribution, some kind of punishment that is fair. We want this because of our sense of fairness and a desire to feel good about the moral universe — that those who violate moral norms will be punished. It is also reasonable and justifiable for us to want more resources than we have, to keep up with the Jones and gain status. Sometimes, as I explained, our desires run out of control. This is especially the case when wanting and liking come

---

[21] Nietzsche, F. *Genealogy of Morality, I,* (D. Smith translation, 1994, Oxford: Oxford University Press); p. 10

unglued, as we saw for the addictions of overeating and overusing of drugs. These are cases where out of control desire leads to out of control action. What I have yet to explain is how a psychology of desire to harm others can spiral out of control and how this psychology connects with and motivates a specific action plan to harm others. Though we may wish to harm another person so that we can feel good, what tips individuals over the threshold from an imagined action to a real action? How does the brain let go of the brakes that typically control our hands and feet, and turn them into weapons that can kill? To begin answering these questions, we look next at people who develop a desire to hurt others — a desire that, for many, is the catalyst that results in excessive harms to innocent victims.

An appetite for violence

Have you ever ruminated on the possibility of hurting someone else? Have you ever harbored a desire to get back at an ex-lover who dissed you or take out a boss who fired you? The following vignettes represent real cases of violent fantasies:

- Prosenjit Poddar told his therapist that he was angry at the woman he was dating because she had expressed interest in other men. He further informed the therapist that he wanted to get a gun and shoot her.

- Carl Carson told friends and doctors about his homicidal thoughts, including a recurring fantasy about his desire to kill people; some of his violent thoughts were triggered by the belief that the government was malicious.

- Seung-Hui Cho wrote a senior college essay in which he described a revenge fantasy, packed with images of retaliation toward those who had what he lacked. He sent his reflections along with excessively violent photographs and videotapes to the New York headquarters of NBC news.

In each vignette, the individual upgraded his violent fantasy to violent reality: Poddar killed the woman he was dating — Tatiana Tarasoff; Carson killed a security guard; and Cho killed 32 people and wounded 25 others in what famously became known as the Virginia Tech massacres. If violent fantasies are closely tied to violent actions, then we may have a burden of responsibility to report the fantasizers to the proper authorities. On this view, therapists would be put in the sticky situation of breaking client confidentiality. After all, if violent fantasies lead to violent actions, then early detection should provide us with greater safety. Or, in legal terms, early detection should provide a screening method to determine *future dangerousness* and thus, our risk of being harmed. These are precisely the issues that Poddar's case triggered. They are issues that led to a controversial California Civil Code decision, often referred to as the Tarasoff duty, that therapists have a responsibility to warn potential victims of danger from a client. This decision, and subsequent interpretations of the law, depend upon the evidence that violent fantasies lead to violent actions.

Studies of individuals who have committed violent acts, including psychopaths, serial murders, and sexual offenders, suggest that violent fantasies are common. For example, records of sexually motivated murders reveal that between 60-80 percent of individuals report having had detailed and recurring fantasies about harming others, some going back to their childhood. Several described running through their fantasies over and over again, deriving pleasure from them, and even going so far as to enact the violence with household objects. Although these results suggest that violent fantasies are common in violent offenders, and may represent precursors to actual violence, they must be treated cautiously given their retroactive quality. In all of these studies, a clinician asked the offender to recall memories of violent fantasies before they committed acts of violence, and sometimes asked about early memories in childhood. We can't be certain, therefore, whether they in fact had these fantasies, believed they did because of the violence they carried out, or fed the clinician what they wanted. And given the correlational nature of the evidence, we certainly can't say whether their fantasies necessarily drove these individuals to commit crimes or were just associated with these events. Cases like Poddar's, Carson's and Cho's are more telling in this light because they reported violent fantasies *before* committing acts of violence.

Several studies of healthy adults without criminal records reveal that between 40-90 percent have had at least one homicidal fantasy. The evolutionary psychologist David Buss reported that 91 percent of men and 84 percent of women had movie quality fantasies about killing someone, including details of the person's identity, the reasons for their desire, and the steps they would take to kill them[22]. Many not only had these fantasies, but started the process of realization in motion. This research indicates that violent fantasies are not the sole province of those with mental health issues or criminal records. Given the numbers, most people that you know have fantasized about killing someone else. Most people you know, however, have not taken the next step and pulled the trigger. These facts minimize the usefulness of using fantasies in a court of law. Nonetheless, we can ask what strengthens the connection between the desire to harm and harm itself? Identifying these connective threads might prove useful in law enforcement, providing protection for victims and help for potential perpetrators.

Normally raised children as young as seven years old are more likely to act aggressively toward their peers if they are self-absorbed in a world of aggressive fantasy.

---

[22] Homicidal fantasies: Duntley, J. D., & Buss, D. M. (2011). Homicide adaptations. *Aggression and Violent Behavior, 16*(5), 399–410.

This correlation between aggression and fantasy is heightened in children who witnessed violence or were subjected to it. Adult men and women are more likely to crave violence after reading an essay about the cathartic power of violent fantasies to flush their aggressive urges than after reading a manifesto against catharsis. Men who engage in aggressive sexual fantasies are more likely to engage in aggression, but only if they are narcissists. Men who engage in deviant sexual fantasies are more likely to enact these fantasies, but only if they exhibit signs of psychopathy. Psychopathy and narcissism are like Siamese twins, inseparable. And yes, the connection between the desire to harm and acts of violence is much stronger in men than women, as it is for the clinical case of psychopathy.

What these studies show is that those who are self-absorbed and play with violent fantasies, are most likely to take these imaginary worlds onto the real world stage, with real harm. They suggest, contrary to the dominant catharsis view among psychoanalysts, that ruminating on violence leads to violence. Like Carlsmith's work on revenge that I described earlier, these results suggest that ruminating on harmful experiences, whether personally experienced or desired in the future, revs up the probability of actually harming someone in the future. Thinking, over and over again, about unsatisfied desires to harm others, is more likely to result in attempts to satisfy these desires. When therapists, especially those influenced by the catharsis view of the mind, encourage their patients to engage in aggressive fantasies to release their pent up energy, we should bring forward malpractice suits as they are accomplices to crime.

We can readily see the connection between the world of fantasy and reality by looking at extreme cases of pathology — cases of desire run wild. This is the world of the lust murderer — individuals with a craving for bizarre and degenerate forms of harming others. This is a world that reveals how an addiction to harming others might arise, tying us back to the discussion of how wanting and liking come apart in the case of food and drugs[23].

---

[23] Lustful violence: Gellerman, D.M, & Suddath, R. (2005). Violent fantasy, dangerousness, and the duty to warn and protect. *Journal of the Academy of Psychiatry and Law, 33*, 484-495; Gerberth. V.J. (1988). Anatomy of a lust murder. *Law and Order Magazine*, 46 (5), pp. 1-6; Gray, N., Watt, A., & Hassan, S. (2003). Behavioral indicators of sadistic sexual murder predict the presence of sadistic sexual fantasy in a normative sample. *Journal of Interpersonal Violence, 18*, 1018-1034; Smith, C.E, Fischer, K.W, & Watson, M.W. (2009). Toward a refined view of aggressive fantasy as a risk factor for aggression: interaction effects involving cognitive and situational variables. *Aggressive. Behavior, 35*(4), 313-323; Crombach, A., Weierstall, R., Schalinski, I., Hecker, T., Ovuga, E., & Elbert, T. (in press). Social status and the desire to resort to violence — a study on former child soldiers of Uganda. *Journal of Aggression, Maltreatment & Trauma 22(55)*; Elbert, T., Weierstall, R., & Schauer, M. (2010). Fascination violence: on mind and brain of man hunters. *European Archives of Psychiatry and Clinical Neuroscience, 260* (Suppl 2), S100-S105; Ruf,

Lust murderers are typically repeat offenders or serial killers. The serial nature of their crimes comes from the fact that they are motivated by recurrent fantasies that create recurrent cravings. They are, effectively, addicted to violence. Their fantasies often entail some kind of paraphilia — an extreme and abnormal sexual arousal to objects, people or situations — played out through some form of sadism — a persistent pattern of sexual or non-sexual pleasure from humiliating, punishing and harming others. Here again we see our uniquely connected and combinatorial mind at work, seamlessly blending pleasure and violence, animate and inanimate attractions, sometimes with benign origins, but often with malignant outcomes; chapter 3 provides the details of how this works and how it evolved.

The paraphilias, like many of the other disorders that appear within the *Diagnostic and Statistical Manual of Mental Health,* fall along a continuum from

M., Schauer, M., Neuner, F., Catani, C., Schauer, E., & Elbert, T. (2010). Narrative exposure therapy for 7- to 16-year-olds: a randomized controlled trial with traumatized refugee children. *Journal of Traumatic Stress, 23*(4), 437-445; Schaal, S., & Elbert, T. (2006). Ten years after the genocide: trauma confrontation and posttraumatic stress in Rwandan adolescents. *Journal of Traumatic Stress, 19*(1), 95-105; Weierstall, R, Schaal, S, Schalinski, I, Dusingizemungu, J, & Elbert, T. (in press). The thrill of being violent as an antidote to posttraumatic stress disorder in Rwandese genocide perpetrators. *Journal of Traumatic Stress*, 1-30.

rather benign forms of voyeurism to erotophono-
philia, the vicious and sadistic killing of an innocent
victim in order to achieve ultimate sexual satisfaction.
Regardless of the particular object or situation driving the paraphilia, individuals develop addictions.
Like other addictions, including those associated
with food, drugs, and alcohol, paraphilic addicts
experience withdrawal. Dangerously for the world
around them, the erotophonophilic or lust killer harbors
sadistic paraphilias, including flagellation – the need
to club, whip or beat someone — anthropophagy —
the desire to eat human body parts — picquerism
— a craving to stab someone or cut off their flesh,
focusing especially on genitals and breasts — and
necrosadism — a yearning to have sexual contact
with the dead. Although these desires may seem
unimaginable, they reveal one facet of the human
mind's potential — a potential that was fully realized
in the mind and actions of Jeffrey Dahmer who flagel-
lated, cannibalized, dismembered, and engaged in
necrophilia with his 17 victims.

Disordered minds, such as those of the lust killer, are
part of the human condition, a continuum that stretches
from individuals who never cache in on their fantasies to
those who not only deliver, but develop — as in addictions to food and drugs — deeper and deeper desires
for harming others without the rewards that come from
such harm. When wanting and liking part company,
with liking falling dormant due to sensitization, wanting
grows in intensity, seeking but failing to find satisfaction.

So begins an appetite for violence, one that can turn into a craving.

Further evidence of the connection between violent fantasies and violent actions, and especially the addictive properties of violence, comes from studies by the psychologist Thomas Elbert who studied child soldiers brainwashed into joining the ranks of the Lord's Resistance Army, Northern Uganda's rebel group. Since its inception in 1987, the LRA has recruited 25-65,000 children, starting with boys and girls as young as 10-12 years old. In detailed interviews and analyses of now retired child soldiers, Elbert discovered that those who had more experience with killing developed stronger, appetite-driven fantasies of killing, a hunger that had to be fulfilled by real killing. As one ex-child soldier noted "The more we killed, the more we acquired a taste for it. If you are allowed to act out this lust it will never let you go again. You could see the lust in our greed popping eyes. [...] It was an unprecedented pleasure for everyone."[24] Not only was the desire to kill converted into killing, but the more they killed, the less they experienced any trauma in later life. Unlike the droves of veterans who have returned from more traditional wars and suffered from post-traumatic stress disorder or PTSD, these child soldiers developed an immunity. For many American war veterans, such as those that returned from Vietnam,

---

[24] Crombach et al. (in press) Social status and the desire to resort to violence — a study on former child soldiers of Uganda. *Journal of Aggression, Maltreatment & Trauma 22(55).*

Afghanistan, and Iraq, there was little interest in killing; many considered war unnecessary, but served to defend their country and national pride. In contrast, the LRA's child soldiers were brain washed into believing that killing was necessary and a sign of importance. Killing that is justified is rewarding, whether the justification is real or the product of self-deception. When self-deception joins the fray — as I further develop in chapter 2 — killing is not only rewarding but virtuous. Killing reaches its most virtuous state in the case of suicide bombers[25].

Work by the economist Alan Krueger reveals that terrorists, including suicide bombers, are neither poor nor uneducated, but typically come from countries that limit civil liberties and focus their attention on democratic societies. From these correlations, Krueger and others have concluded that individuals living in societies that constrain human freedom, and who desire the pub- lic attention of large powerful democratic nations, are most at risk of terrorism. As the anthropologist Scott Atran correctly notes, however, a closer look at the characteristics of these terrorists, and the countries from which they emanate, reveals an important twist to this account. Al-Qaeda, for example, consists of members that come from a wide variety of countries, some poor

---

[25] Terrorism and the suicide bomber: Atran, S. (2009). Who becomes a terrorist today? *Perspectives on Terrorism*, 3(5), 1–9; Krueger, A. (2007). *What Makes a Terrorist: Economics and the Roots of Terrorism*. (Princeton, N.J.: Princeton University Press).

and some not, some democratic and some authoritarian; Hezbollah and Hamas, in contrast, consist of a more circumspect membership. Looking to Europe, a high incidence of terrorist activities were ignited by individuals with unconstrained civil liberties. In a more focused set of analyses, Atran found that among the wealthiest and most educated Palestinians, there was more support for suicide attacks, whereas close to 80% of terrorist attacks by Jemaah Islamiya — a southeastern ally of Al-Qaeda — were carried out by uneducated, unskilled individuals. These analyses suggest that education, poverty, and political ideology are not predictors of terrorist activity. Rather, as Atran's studies reveal, the best predictors of terrorism are small social networks, including the people they play soccer with, marry, and befriend in their neighborhood. When members of ones most intimate social network promote violent ideology and convey the excitement of public attention, the odds of joining in on a violent mission rise, even if this means self-sacrifice. As Atran notes, the 21st century of terrorism reveals the power of "adventure, dreams of glory, and esteem in the eyes of peers" (p.8) to fuel excessive acts of harm by young people.

Given the material discussed thus far, one plausible conclusion is that the potentially addictive nature of violence can only be seen in relatively uncommon and extreme cases: lust murderers, brain washed child soldiers and suicide bombers. Though these cases provide vivid illustrations of how unsatisfied desires can drive excessive violence, other, far more common examples surround us.

One of the most common contexts is domestic or intimate partner violence[26]. In the United States, an estimated 1.3 million women per year are battered by their male partners, with the number of attacks within a couple increasing both when women attempt to leave or when they report the violence to the authorities. This pattern suggests that partner withdrawal, just like substance abuse withdrawal, causes an increasing desire to harm, leading in many cases to homicide. In the United States, the Bureau of Justice Statistics reported that there were 1640 females killed by their partners in 2007. Based on these patterns, a number of clinicians have noted the parallels between domestic violence and addictive disorders. As discussed by Richard Irons and Jennifer Schneider, when domestic violence erupts, it, like an addiction to drugs or alcohol, is characterized by a loss of self-control, repeated actions with adverse consequences, obsessive thoughts, denial of the problem, and desensitization, with violence no longer satisfying the desire to harm. Domestic violence is rarely a one time affair. Domestic violence grows from an unsatisfied desire to control one's partner. But like the consumption of alcohol, violence itself fails to satisfy the desire because wanting and liking have parted company. Violence becomes addictive.

---

[26] Domestic violence as an addiction: Irons, R. & Schneider, J.P. (1997). When is domestic violence a hidden face of addiction? *Journal of Psychoactive Drugs* 29(4): 337-344; Wekerle, C. & Wall, A-M. (2002). *The Violence and Addiction Equation*. New York, Routledge.

Lust murderers, child soldiers, suicide bombers, and batterers in an abusive relationship move easily from violent desires to violent actions. For all, harming others is addictive. For all, the addiction is aided by a psychology of denial. Lust murderers fuel their appetite for violence by thinking of their victims as objects, child soldiers do it by means of self-deception, suicide bombers by the belief in a just cause, and battering spouses by creating the false belief that they deserve control. Objectification, self-deception, and ideological justification are forms of denial that loosen the grip of our moral sense, and help complete the recipe for evil.

Marc D. Hauser

Recommended books

Atran, S. (2010). *Talking to the Enemy: Faith, Brotherhood, and the (Un)Making of Terrorists.* New York: Harper Collins.

Bloom, P. (2010). *How Pleasure Works.* New York: W.W. Norton.

Krueger, A. (2007). *What Makes a Terrorist: Economics and the Roots of Terrorism.* Princeton, N : Princeton University Press.

Staub, E. (2010). *Overcoming Evil.* New York: Oxford University Press.

two: ravages of denial

Self denial is not a virtue: it is only the effect of prudence on rascality.

– George Bernard Shaw

The still life and tableau vivant represent two art forms that involve careful placement of objects into specific positions. The artist, using canvas or camera, recreates a staged scene. In a still life, the focus is typically on inanimate objects such as pitchers, books, or fruit, while in a tableau vivant the focus is on animate objects, including humans and other animals.

The tableau vivant reached new heights when Corporal Charles Graner decided to use his photographic talents to pose Iraqi detainees at Abu Ghraib into human pyramids, hooded scarecrows attached to electrical wires, dogs on leashes, and piles of naked flesh. Unlike the original art form which involved wealthy guests at dinner parties who voluntarily put on costumes and assumed silly poses, the Iraqi detainees were forced into humiliating positions, dehumanizing them into inanimate objects or animals. This represents a massive distortion of reality, but is not the representation of a delusional mind. Graner believed that the only way to protect his group was to crush and humiliate the other group's spirit, and the only way to do that was to dehumanize them. For those of us who dared to stare at these pictures, or saw them come to life in Errol Morris's superb film *Standard Operating Procedure,* they provide a wake-up call to the horrific ways in which we recruit dehumanization and self-deception to deny the moral fabric of other humans. This chapter explains how these two components of denial work, how they are enlisted to satisfy our desires, and often lead to the destruction of innocent lives.

Attempting to satisfy our desires frequently brings us into opposition with others who are interested in the same resources, as well as moral sanctions that prohibit particular actions. When moral constraints operate, either as intuitively understood norms or explicitly recognized laws, they set guidelines for what is right or wrong, what is praiseworthy or blameworthy, and who counts within the circle of morally relevant individuals. When we dehumanize another human being, we have taken them out of the circle of moral consideration, thereby lifting a significant constraint on our desire to procure resources. When we self-deceive, we justify to ourselves and often to others that new moral norms are necessary to address a societal problem. With new norms in place, actions that were prohibited under the old regime are not only legitimate, but sanctioned. So begins a process of denial that enables us to satisfy our desires.

Before I explain how we dehumanize and self-deceive, I need to explain the process of humanization, the capacity to perceive some things but not others with human qualities and moral worth. This is an important process as it shapes our perception of who can cause excessive harm and who can suffer from it. Rocks can cause great pain — as in landslides — but we don't hold them responsible for the harm caused because they lack intentions, beliefs, goals, and desires. Rocks can also be crushed, pulverized into sand by humans working in a quarry. But rocks are neither innocent nor victims as they have no moral worth, no capacity to suffer, and no ability to intentionally harm another. If not rocks, what? If dehumanization is taking away human qualities and moral worth, who has these to take away?

iHuman

Aristotle developed the distinction between moral agents and moral patients. Agents have responsibility for others' well being, whereas patients deserve moral consideration and care. Moral agents are potential evildoers who can cause excessive harm to moral patients, but not the other way around. Moral patients may well cause harm, but they lack the cognitive wherewithal to both reflect upon the moral consequences of their actions and the reasons why certain actions are morally forbidden.

Marc D. Hauser

Aristotle's distinction between moral agents and patients raises fundamental questions concerning the features we use to distinguish them[27]. What enables

---

[27] The features that guide our perception of humanness: Bastian, B., Laham, S.M, Wilson, S., Haslam, N., & Koval, P. (2011). Blaming, praising, and protecting our humanity: The implications of everyday dehumanization for judgments of moral status. *British Journal of Social Psychology, 50*(3), 469-483; Graham, J., & Haidt, J. (2011). Sacred values and evil adversaries: a moral foundations approach. In: M. Mikulincer & P.R. Shaver (Eds.), *The Social Psychology of Morality*; pp. 11-32, Washington, DC: American Psychological Association; Gray, H.M., Gray, K., & Wegner, D.M. (2007). Dimensions of mind perception. *Science, 315*(5812), 619-619; Gray, K, & Wegner, D M. (2011). Morality takes two: dyadic morality and mind perception. In: M. Mikulincer & P.R. Shaver (Eds.), *The Social Psychology of Morality*, pp. 109-128, Washington, DC: American Psychological Association; Gray, K., Jenkins, A., Heberlein, A. S., & Wegner, D. M. (2011). Distortions of mind perception in psychopathology. *Proceedings of the National Academy of Sciences (USA)*, 108: 477–479; Haslam, N., Bastian, B., Laham, S., & Lougham, S. (2011). Humanness, dehumanization, and moral psychology. In: M. Mikulincer & P.R. Shaver (Eds.), *The Social Psychology of Morality*, pp. 203-218, Washington, DC: American Psychological Association; Loughnan, S., Leidner, B., Doron, G., Haslam, N., Kashima, Y., Tong, J., & Yeung, V. (2010). Universal biases in self-perception: Better and more human than average. *British Journal of Social Psychology, 49*(3), 627-636; Martinez, Piff, P., Mendoza-Denton, R., &

someone to have responsibility and know when to deploy it in the service of helping another? What capacities make an individual worthy of a moral agent's care, including the delivery of help and the avoidance of harm? The psychologists Kurt Gray and Daniel Wegner addressed this problem in a series of studies. In one experiment, a large internet population of adults compared the qualities of different things, including humans at different stages of development (fetus, baby, child, and adult), an adult human in a vegetative state, a dead human, nonhuman animals (frog, pet dog, chimpanzee), God, and a socially savvy robot. Subjects judged different pairings of these *things* on a wide range of dimensions, including which was more likely to develop a unique personality, feel embarrassed, suffer pain, distinguish right from wrong, experience conscious awareness, exert greater self-control, plan ahead, develop fears, feel pleasure, and erupt into rage. Subjects also provided their personal opinions on which individual, within the pair, they liked most, wanted to make happy or destroy, was most deserving of punishment, and most likely had a soul.

Presumably, everyone reading about the design of this study has already formed an opinion about some of the comparisons. Presumably, everyone believes that a living adult is more consciously aware than a dead person, fetus, dog, and robot. Presumably,

Hinshaw, S. P. (2011). The power of a label: Mental illness diagnoses, ascribed humanity, and social Rejection. *Journal of Social and Clinical Psychology, 30*(1): 1–23.

everyone believes that all animals feel more pain than a dead human or a robot. And presumably, everyone would rather make a dog happy than a frog, and would be more likely to allocate souls to fetuses, babies, and adult humans than to robots and frogs. But are we more or less conscious than God? Does a chimpanzee feel more embarrassed than a baby? Can a person in a vegetative state feel more pleasure than a frog or robot? What dimensions, if any, cause us to lasso some things together but not others? What things cluster together and why?

Gray and Wegner added up the responses and produced a landscape defined by two dimensions: experience and agency. Experience included properties such as hunger, fear, pain, pleasure, rage, desire, consciousness, pride, embarrassment, and joy. Agency included self-control, morality, memory, emotion recognition, planning, communication and thinking. Experience aligned with feelings, agency with thinking. With these dimensions, we find God at one edge, high in agency and low in experience. On the opposite side, huddled together on the landscape defined by low agency and high experience, we find fetuses, frogs, and people in a vegetative state. Clustered inside the high agency and experience space we find adult men and women, whereas robots and dead people land in the low experience and middling agency space; dogs, chimpanzees, and human kids are clustered together in the high experience and mid-level agency.

In addition to classification, the dimensions of experience and agency also play an active role in guiding individuals' judgments to punish, provide pleasure, and avoid harm. The simple rule of thumb is: avoid harming things high in experience and punish things high in agency. Even young children understand that you can kick a rock but not a dog, and that you can punish dogs but not rocks.

What this work suggests is that people have strong intuitions about which things are morally responsible as agents and which deserve our moral concerns as patients. Moral patients are high in experience and can thus suffer as victims, innocent or not. This is why many countries have created laws against harming nonhuman animals, including restrictions on which animals can serve in laboratory experiments, what can be done to them, and how they should be housed. This is also why we don't do experiments on fetuses, newborns, adults in a vegetative state, or humans with neurological disorders that eliminate aspects of their experience and agency; when people carry out such experiments, as in the case of the Nazi doctors, those responsible are perceived by many as inhumane moral monsters.

Once something enters the arena of moral patients, we tend to leave them within this space even if they lose particular capacities; for example, when human adults lose self-control or compassion due to brain injury or a developmental disorder, this has no bearing on their status as moral patients. Conversely, if scientists discover that an animal outside the arena of moral patient-hood has capacities of experience and agency that are on a par with those inside, this evidence often promotes their legal status and protection. Such was the fate of the octopus, an invertebrate once classified by Aristotle as stupid, but today elevated to the company of much smarter animals such as chimpanzees and dolphins who solve novel problems, deploy trickery, and show some evidence of being aware of their behavior. As such, they are one of the few invertebrates to enjoy legal protection and thus, specific forms of care when they are kept in captivity. The octopus is a moral patient.

Moral agents are an altogether different species. They are high in agency, meaning they can distinguish right from wrong, exert self-control in the context of selfish temptation, can be blamed and punished, and are expected to care for moral patients. Moral agents are also high in experience, especially the capacity to feel pain and recognize pain and suffering in others. These differences place a burden of responsibility on moral agents that is absent from moral patients.

Moral agents and patients have moral worth. But as with all entities that have worth or value, some are more valuable than others. So it is with moral worth. This is where departures from humanness get interesting, dangerously so. When we strip individuals of their moral worth, denying them qualities that define humanness, we enter a world of denial that can energize our desires and justify excessive harm.

Moral zeroes

Chances are that you believe you are more compassionate, ethical, and rational than most people you know. My presumptuous guess is based on the evidence from several studies showing that individuals consider themselves to be more human — as defined by the dimensions of experience and agency — and to have greater moral worth than other individuals. When individuals are socially ostracized and excluded from a group, they judge themselves as less human, and so do the spectators who watch the ostracism unfold. Individuals judge members of their own group to be more human and morally worthy than those outside the group, no matter how small or broad the group is. What counts is our overall sense of how we compare to others and the dimensions used to calibrate this similarity metric. Understanding this process is a key step in explaining how we can deceive ourselves into believing that another human being is morally worthless, and thus worthy of exclusion from the circle of moral patients. Once they are displaced outside, causing them harm either feels justifiable because they are dangerous predators or parasites, or there are no feelings at all because they are objects — moral zeroes.

The social psychologist Nick Haslam carried out several experiments to determine how our rating of a group's humanness influences how much we praise, blame, and protect them, as well as whether we believe that rehabilitation or punishment is most appropriate after they have done something wrong[28]. Haslam based his study on the idea, supported by law, science, and folk intuitions that we blame, praise, and punish only those who do bad things on purpose as opposed to by accident. Conversely, we favor rehabilitation in those cases where we believe that the person can right a wrong, learning a lesson from a prior transgression. For example, we don't blame a five year old for picking up a gun and shooting someone, and nor do we sentence him to life in prison. We blame his caretakers and find ways to help the child understand the consequences of his actions. At some point, usually around the eighteenth birthday in most countries, we blame the shooter, holding him responsible and punishing him, often with life in prison or the termination of life itself.

Haslam's subjects started by rating several different social groups along different dimensions of humanness. Though the dimensions were slightly different from those used by Gray and Wegner, they generally corresponded to experience and agency, including compassion, warmth, and a sense of community on the one hand, and reason, self-control, civility, and refinement on the other. The target social groups were associated with negative or positive stereotypes such as the homeless,

---

[28] For a review of dehumanization, see Haslam, N., Bastian, B., Laham, S., & Lougham, S. (2011). Humanness, dehumanization, and moral psychology. In M. Mikulincer & P.R. Shaver (Eds.), *The Social Psychology of Morality*, pp. 203-218, Washington, DC, American Psychological Association.

mentally disabled, athletes, politicians, doctors, lawyers, gays, and different religious groups. Next subjects imagined that a member of one of these groups had acted morally or immorally, or had been mistreated in some way. Then they decided whether the person should be praised for a particular moral act such as returning a wallet, considered responsible for an immoral act such as breaking a promise, helped out for mistreatment such as being pushed out of line by a person in a hurry, and punished or rehabilitated for wrongful behavior.

Haslam's results generated a landscape of humanness very much like Gray and Wegner's. Those groups rated highly in terms of agency were more likely to be blamed and punished. Those groups rated high in experience were more likely to be praised, protected, and placed into rehabilitation. Those groups perceived as more emotional, compassionate and warm — components of experience — were praised more, whereas those perceived as more civil and rational — components of agency — were praised less. Overall, the more a group tilts toward the experience end of the spectrum, the more we see them as moral patients, deserving of our care and compassion. The more a group tilts toward the agency end of the spectrum, the more we see them as moral agents, having responsibilities and duties to act morally. When one group perceives another as lacking in experience and agency, they are perceived as moral zeroes. As in mathematics, subtracting zero from something leaves the mathematical universe unchanged. So it is with subtracting moral zeroes, at least from the perspective of those doing the subtracting. How a group shifts their perception of another's moral worth and thus, shifts their sense of which actions are morally justified, requires further explanation as it represents a fundamental enabler of evildoing.

When we lower our sense of another's value, we are willing to violate our sense of the sacred, engaging in trade-offs that are normally taboo. Experiments by the psychologists Philip Tetlock and Jonathan Haidt help us see what is sacred by asking individuals what they would pay to do something sacrilegious[29]. If something is sacred, of great moral worth either personally or to your group, could you be paid off by a wealthy investor to give up the object or engage in an act against it? For each of the acts below, think about your payoff point in dollars from $0 (for free) to $1million, including the option of saying that you would never do it for any amount of money. Keep in mind that if you choose to carry out an act and receive payment you will not suffer any consequences:

---

[29] Being paid off to do something wrong: Graham, J. & Haidt, J. (2011). Sacred values and evil adversaries: A moral foundations approach. In: *The Social Psychology of Morality: Exploring the causes of good and evil*. P. Shaver & M. Mikulincer (Eds.), Washington, DC: American Psychological Association; Tetlock, P. E., Kristel, O., Elson, B., Green, M., & Lerner, J. (2000). The psychology of the unthinkable: Taboo trade-offs, forbidden base rates, and heretical counterfactuals. *Journal of Personality and Social Psychology, 78,* 853–870.

- Kick a dog in the head, hard.
- Sign a secret but binding pledge to hire only people of your race into your company.
- Burn your country's flag in private.
- Throw a rotten tomato at a political leader that you dislike.
- Get a pint transfusion of disease-free, compatible blood from a convicted child molester.

If you are like the subjects in these experiments, the mere process of considering a payoff, even for a short period of time, will have turned your stomach into knots and triggered a deep sense of disgust. This is because violating the sacred is akin to violating our sense of humanness. It is playing with the devil, accepting a Faustian offer of money to strip something of its moral worth. As Haidt notes, even though it is sacrilege to accept payment across different moral concerns, including avoiding harm, acting fairly, and respecting authority, different experiences can modulate the aversion we feel when we imagine such transgressions. Women typically demand more money for each of these acts than men, and more often reject them as taboo. Those who lean toward the conservative end of the political spectrum either ask for more money or consider the act taboo when compared to liberals, and this is especially the case for questions focused on acting against an in-group (race), an authority figure (political leader), or one's purity (blood transfusion). What this suggests is that certain experiences can distort what we consider morally worthy or sacred. It suggests that we can flip our values in the face of tempting alternatives. It suggests that we can be tempted to treat others as moral zeroes.

The scientific evidence presented in this section reveals that our decisions to treat others according to different moral principles or norms is powerfully affected by our sense of what counts as another human being. What counts includes at least two important dimensions, one focused on agency and the other on experience. These dimensions determine whether we blame or praise someone, punish or rehabilitate them, and ultimately, include or exclude them from the inner circles of moral agents or moral patients. Those who fall outside these two inner circles are morally worthless. Those who are morally worthless can be destroyed or banished. Some things are justifiably excluded and fit with our general sense of reality — rocks, dirt, cardboard boxes, plastic balls, and pieces of glass. Other things are excluded because they don't fit with our values of what reality should be. This is where distortion and denial enter the process. This is where we create walls around members of one group in order to keep others out. This is where we express partiality instead of the impartiality that Lady Justice champions with her two balanced scales and blindfold-covered eyes. This is where we exclude others from our inner sanctum in order to justify great harms. How is the inner sanctum set up and put into action over a lifetime, sometimes for legitimate causes and sometimes for illegitimate and unconscionable causes?

Populating the inner sanctum

Elie Wiesel, the Nobel Peace Laureate and Holocaust survivor, remarked that "anti-Semitism is the oldest group prejudice in history."[30] This claim may well be true of human written history, but is most definitely false if one considers the fact that all social animals and human societies, including the hunter-gatherer groups that are descendants of people that predated the Jews, hate some individuals and love others. Prejudice, though often based on deep seated ideological biases and stereotypes that humans invent, is, at root, a form of partiality. Every social animal expresses partiality. This is a highly adaptive and ancient psychology, promoting the care of young, investment in mates, and escape strategies against enemies. Humans are no different, except for the role that our unique brains play in fueling partiality with ideology and symbolism. Sometimes when we express our partiality it is for the noble cause of caring for our children and for defending ideological beliefs surrounding humanitarian causes, including defense of basic human rights. Sometimes when we express our partiality, it is for the ignoble cause of destroying others. In this section, I explain how human partiality works, from its earliest incarnation in infancy to its most articulated and culturally developed form in adult-

---

[30] This is a comment Elie Wiesel has made repeatedly in his writings and speeches, reproduced here: http://www.thirteen.org/openmind/history/anti-semitism-redux/1609/

hood[31]. At birth, newborns prefer to listen to people speaking their native language over those speaking another language. Soon thereafter, infants prefer to listen to their native dialect over a non-native dialect, and look longer at people from their own race than those from another race. This suggests that prior to any significant cultural indoctrination, infants can discriminate between different languages, dialects, and racial groups — all markers of group identity. But do they care about these differences? Do they form social preferences based on these distinctions? Would

---

[31] Populating the inner sanctum: Cosmides, L., Tooby, J., & Kurzban, R. (2003). Perceptions of race. *Trends in Cognitive Science, 7*(4), 173-179; Kinzler, K.D, Corriveau, K.H, & Harris, P.L. (2011). Children's selective trust in native-accented speakers. *Developmental Science, 14*(1), 106-111; Kinzler, K.D, & Spelke, E.S. (2011). Do infants show social preferences for people differing in race? *Cognition, 119*(1), 1-9; Kinzler, K.D., Dupoux, E., & Spelke, E.S. (2007). The native language of social cognition. *Proceedings of the National Academy of Sciences (USA), 104*(30), 12577-12580; Kinzler, K.D, Shutts, K., Dejesus, J., & Spelke, E.S. (2009). Accent trumps race in guiding children's social preferences. *Social Cognition, 27*(4), 623-634; Kurzban, R, Tooby, J., & Cosmides, L. (2001). Can race be erased? Coalitional computation and social categorization. *Proceedings of the National Academy of Sciences (USA), 98*(26), 15387-15392; Markus, H.R. (2008) Pride, prejudice, and ambivalence: toward a unified theory of race and ethnicity. *American Psychologist, 63 (8),* 651-70.

a young baby prefer to take a toy from an unfamiliar person who speaks the same or different language, from the same or different race? To answer these questions, the developmental psychologist Katharine Kinzler gave five months old babies a test.

Babies born into families of one race and one language sat on their mother's lap in front of two monitors, each presenting short video clips of different people. After watching the videos, Kinzler created a bit of magic. The people in the monitor appeared to emerge from the image and offer the baby a toy. The trick: a real person, hidden beneath the monitor, synchronized her reach with the reach in the monitor. Who would the baby choose given that both people offered the same toy? Babies grabbed the toy from people speaking the native over non-native language and the native-accent over the non-native accent, but showed no preference for the native race. Thus, early in life the connection between discrimination and social preference is well established for language, but not for race. When do things change for race?

Kinzler carried out another series of experiments on race with one group of two and a half-year old children and a second group of five-year olds. Though these two age groups required different methods, both focused on the child's preferences, including who they would share toys with and who they would prefer as friends. The two and a half year olds showed no preferences, whereas the five year olds preferred their own race. Race is therefore a slowly developing category, at least in terms of its impact on social preferences, and especially when contrasted with both language and accent.

Kinzler took these studies one step further to ask: What's more important to a young child building an inner sanctum of trusted others — race or language? Would they rather interact with someone of the same race who speaks a foreign language or someone of a different race who speaks the native language? Using similar procedures, Kinzler showed that by five years of age, language trumps race. Children would rather interact with someone from a different race speaking the same language than someone of the same race speaking a foreign language.

Why would language trump race? Kinzler's answer relies on an idea developed by the evolutionary psychologist Robert Kurzban. Imagine a hunter-gatherer in South Africa, living during the earliest stages of our evolutionary history. Consider the fact that this individual was constantly in search of food to eat, places to sleep, and water to drink. Consider further that this individual had to compete with others to satisfy his desire for such resources. As soon as this individual moved outside of his primary living area, he met strangers. All were from the same race given that racial differences didn't emerge until relatively late in human evolution, well after our ancestors took their first steps out of Africa. If we are thinking about an evolved psychology for bonding with members of our own group and fighting those outside, our ancestors would have been blind to race as it was not yet an emergent property of our species. Language was, however, a property of our species, one that varied across populations. A hunter-gatherer walking the plains of South Africa would indeed have run into people speaking either a completely different language, or the same language with a different dialect. New languages are not easily acquired and nor are new dialects. It takes talent to speak a new language or dialect without a trace of ones origins.

The babies in Kinzler's experiments tell us something important: race and language are both discriminable from an early age, but language precedes race in terms of its prejudicing effects on infants' social preferences. Language trumps race because it is a better predictor of membership within the inner sanctum, at least early in life. Ultimately, both language and race are used to close off some from the inner sanctum and allow others in. But this is only the beginning of our prejudicial outlook. With time and experience, we not only discriminate against others using finer and finer distinctions, but use our discriminating powers to close more doors. This is a process that I discussed in the context of Fehr's research showing that children are more likely to share candy with kids from the same school than with kids from a different school. This is a process that grows, creating rich cooperative networks within communities — think religious groups — and equally rich antagonistic networks with those outside the community — again, think religious groups.

Closed doors

As adults, we tend to rely on rules of thumb to guide our social interactions, including who we trust and who we distrust. We tend to trust those we know more than those we don't know. Within the circle of those we know, we believe those who are more like us than those who are unlike us, using unchangeable parts of the body (race, height, eye color), flexible psychological features (food preferences, sports' interests, religious beliefs), and features that vary within a constrained space of possibilities (language and intelligence). Together, these different dimensions cause us to close the door on some and open it to others. This makes sense from an evolutionary per-spective, a point that I briefly made in the last section and that I will explain in greater detail in chapter 3: in a world with limited resources, there is strong pres-sure to help our kin and friends within the group, and to keep strangers at bay. This ancient capacity will always be with us, despite our attempts at equality and impartiality.

Consider language again. If you can't understand someone because they speak a foreign language or because their accent in the native language is too heavy, then the issue is not trust, but comprehension. But what if you can understand the person perfectly well, but they speak with a foreign accent, either one from a different country (such as a Frenchman speaking English) or one from the same country but a different region (such as a Southerner from Georgia asking a New Yorker for directions)? [32]

Subjects in an experiment first listened to people reading trivia, such as "A giraffe can go without water longer than a camel can," and then judged whether the sentence was true or false. If the sentence was read in a foreign accent, subjects were more likely to say that it was false than if it was read in the native accent. Subjects voiced this opinion even though the experimenter told them that the reader was not expressing an opinion, but merely reading the passage as instructed. In a second experiment, British subjects listening to a non-guilty plea by a person on trial were more likely to judge the person as guilty if he committed a blue collar crime and spoke with a non-standard British accent (e.g., Australian). In contrast, they were more likely to judge a white collar criminal as guilty if he spoke with a standard British accent. Even within the class of British accents, biases emerged: subjects from the Worcester region were more likely to judge supposed criminals as guilty if they spoke with a Birmingham accent than with a

---

[32] The prejudicing power of language: Dixon, J, Mahoney, B, & Cocks, R. (2002). Effects of regional accent, race, and crime type on attributions of guilt. *Journal of Language and Social Psychology, 21*(2), 162-168; Lev-Ari, S., & Keysar, B. (2010). Why don't we believe non-native speakers? The influence of accent on credibility. *Journal of Experimental Social Psychology, 46*(6), 1093-1096.

Marc D. Hauser

Worcester accent. Together, these studies paint a bleak picture: accents from an out-group are perceived as less truthful than others, and in the context of a criminal case, more guilty as well.

Accents are learned early in life. Once in place, our accents are both clear markers of our origins and difficult to undue. As such, they are reliable indicators of at least one dimension of group membership. What about dimensions that can readily be acquired at any point in life and just as easily dropped? How do these influence not only our perception of those who share these dimensions in common, but how we treat them?[33]

---

[33] Unconscious attitudes and prejudice: Avenanti, A., Sirigu, A., & Aglioti, S.M. (2010). Racial bias reduces empathic sensorimotor resonance with other-race pain. *Current Biology, 20*, 1018-1022; Balas, B., Westerlund, A., Hung, K., & Nelson, C.A. (2011). Shape, color and the other-race effect in the infant brain. *Developmental Science, 14*(4), 892-900; Banaji, M.R. (2001). Implicit attitudes can be measured. In: H.L. Roediger, J.S. Nairne, I. Neath & A. Surprenant (Eds.), *The Nature of Remembering: Essays in Honor of Robert G. Crowder.* Washington: American Psychological Association; Correll, J, Wittenbrink, B, Park, B, Judd, C.M, & Goyle, A. (2010). Dangerous enough: Moderating racial bias with contextual threat cues. *Journal of Experimental Social Psychology, 47*, 184-189; Chiao, J.Y., & Mathur, V.A. (2010). Intergroup empathy: How does race affect empathic neural responses? *Current Biology, 20*(11), R478-R480; Dunham, Y., Baron, A., & Banaji, M. (2008). The development of implicit intergroup cognition.

In the last chapter I discussed a study by Tania Singer in which both men and women showed more pain empathy — as revealed by activation in the insula region of the brain — when they watched a cooperator experiencing pain. Men also showed a reduction of activity in this area when a cheater experienced pain, and increased activity in a reward area — the nucleus accumbens. We feel compassion for those who cooperate with us, as cooperation is a sign of group solidarity and membership. We lack these feelings toward cheaters because they are either competitors outside of our group or individuals inside who don't deserve to be. When a cheater has

---

Trends in Cognitive Sciences, 12(7), 248-253; Gutsell, J.N., & Inzlicht, M. (2010). Empathy constrained: Prejudice predicts reduced mental simulation of actions during observation of outgroups. *Journal of Experimental Social Psychology, 46*, 841-845; Hein, G., Silani, G., Preuschoff, K., & Batson, C.D. (2010). Neural responses to ingroup and outgroup members' suffering predict individual differences in costly helping. *Neuron, 68*, 149-160; Ito, T.A, & Bartholow, B.D. (2009). The neural correlates of race. *Trends in Cognitive Sciences, 13*(12), 524-531. Masten, C.L, Telzer, E.H, & Eisenberger, N.I. (2011). An FMRI investigation of attributing negative social treatment to racial discrimination. *Journal of Cognitive Neuroscience, 23*(5), 1042-1051; Xu, X., Zuo, X., Wang, X., & Han, S.. (2009). Do you feel my pain? Racial group membership modulates empathic neural responses. *Journal of Neuroscience, 29*(26), 8525-8529.

been caught and punished for his crime, we rejoice, feeling schadenfreude for just deserts.

Singer took this work further, asking whether an individual's support for a sport's team — a dimension that can be acquired and dropped at will — might similarly modulate both the feeling of pain empathy as well as reward. Whether playing on a team or supporting them as a fan, we identify the players as members of our group, and those on other teams as members of an out-group. Subjects in the experiment — all soccer fanatics — sat in a scanner and watched as a player from their favorite team or a rival experienced pain. Next, Singer provided subjects with three options for interacting with these players: help them by personally taking on some of the pain they would receive, let them take on all the pain but watch a video as distraction, or let them take on all the pain and watch as it happens. Option one is costly altruism, two is blissful ignorance, and three is schadenfreude.

In parallel with the earlier work on fairness, here too Singer observed greater pain empathy when the favorite team player experienced pain than when the rival experienced pain. She also observed that subjects were more likely to help favorite team players by taking on some of their pain, but more likely to watch rivals receive pain. The higher the activation level in the insula, the more they took on their favorite team player's pain session — the more they helped. When they watched rivals experience pain, there was significant activation in the nucleus accumbens. They felt an immediate honey hit, joy over the rival's pain. The higher the activation in this reward area, the more likely they were to choose the option of watching the rival experience pain — like watching a public execution and cheering for just deserts.

Singer's results suggest that individual differences in our compassion toward others in pain predicts our willingness to help them. It reveals another dimension, like language, that biases our sense of justice, both in our judgments and in our behavior. Conversely, individual differences in our joy over others' misery predicts our willingness to allow others to suffer, suppress our instincts to help and, I suggest, facilitate our capacity to harm.

Unlike our preferences for sports teams, race is a feature of group membership that is fixed at birth. But like sports teams, our perception of race biases our judgments and actions. As noted earlier, babies stare longer at faces of people from the same race than faces of people from a different race and by the pre-school years, are more likely to show social preferences for peers and adults of the same race. In brain imaging studies, specific areas activate when we process faces as opposed to other objects, and one tenth of a second later, other associated regions activate when we process race. This rapid activation occurs whether we are consciously engaged in classifying faces by race or not; for example, the same areas activate even when we are forced to focus on gender or familiarity. This suggests that from the brain's perspective, we don't have an option of processing a person's race. The brain automatically and unconsciously hands us this information, like it or not. And once we are handed this information, it can influence how we treat others, including our compassion for them. This point is supported by several studies.

In one study, modeled after Singer's experiment on cooperators and cheaters, Caucasian, Black and Asian subjects sat in a scanner and observed other individuals experience a painful event. Subjects showed stronger activation in the pain-related areas of the brain when viewing individuals from the same race experience pain than when viewing individuals of a different race. In a second study, subjects sitting in a brain scanner played a computerized game involving social ostracism. Black subjects showed stronger activation in areas of the brain involved in social pain when excluded by Caucasian players than when excluded by Black players. Together, these studies suggest that when others suffer and we have the opportunity to help them, we are more likely to help those of the same race, and feel good about it. Conversely, when others are unlike us, we are more than happy to let them suffer or help the cause by causing greater suffering.

Our biases, both unconscious and conscious, influence our compassion toward others and our motivation to help. This statement is true whether we are looking at evidence from young children or adults, and using measures that assess activity patterns in the brain, sensory perception, or behavioral judgment. Beginning with an evolutionarily ancient brain system that was designed to distinguish friendly in-group members and antagonistic out-group members, we populate the inner sanctum with people who we perceive as most like us, using both fixed and variable features. With time, the walls surrounding this sanctum close, attributing the full richness of human nature to those within and bleaching it from those outside.

Bleaching humanity

Draw an imaginary circle around yourself with a diameter of about fifty feet. Now imagine packing this circle with people, forming an expanding set of concentric circles that radiates out from those closest to you to those you know less well. Based on analyses by the sociologists Nicholas Christakis and James Fowler, and summarized in their book *Connected*, the majority of people within the inner circles will be like you in a number of ways, including their race, religion, political affiliation, food preferences, and aesthetics. This includes family members and close friends, but also those we work with, vote for, and play with. As you travel out from the inner core, you will find less in common. Some in the outer circles will not only have less in common, but as noted earlier in this chapter, will be perceived as less human, stripped of dimensions of experience and agency that define humanity.

When we strip others of their humanity — a process that can occur both consciously and unconsciously — we may perceive them as either inanimate objects or as animals. Those we perceive as objects lack core

Marc D. Hauser

aspects of human nature, including emotional sensitivity, warmth, and flexibility. Those we perceive as animals lack uniquely human qualities such as rationality, self-control, moral sensibility, and civility. Of those we see as animals, some will seem like kin to the domesticated form and thus controllable as property; others will seem like wild animals and thus dangerous, dirty and deserving of elimination. However we engage this process, we have bleached individuals of their humanity. This process, one that occurs in both everyday life and in cases of conflict, has allowed us to treat the mentally and physically disabled like animals, to consider women as sexual property, justify slavery, deny certain races the opportunity to vote and receive education, and mandate ethnic cleansing. Dehumanization is a form of denial that, I suggest, enables gratuitous cruelty.

To set the stage for the scientific evidence on dehumanization[34], there are two important points to

---

[34] Bleaching humanity of its essence: Cheng, Y., Lin, C.P., Liu, H.L., Hsu, Y.Y., Lim, K.E., Hung, D., & Decety, J. (2007). Expertise modulates the perception of pain in others. *Current Biology, 17*(19), 1708-1713; Decety, J. (2009). Empathy, sympathy and the perception of pain. *Pain, 145*(3), 365-366; Decety, J., Michalska, K.J., Akitsuki, Y., & Lahey, B.B. (2009). Atypical empathic responses in adolescents with aggressive conduct disorder: a functional MRI investigation. *Biological Psychology, 80*(2), 203-211; Jackson, P., Meltzoff, A., & Decety, J. (2005). How do we perceive the pain of others? A window into the neural processes involved in empathy. *Neuroimage,*

keep in mind. The first is that dehumanization is not restricted to a particular period of time in history nor to a particular group of people. As recently reviewed by the philosopher David Livingstone Smith in his book *Less than Human*, in every major recorded instance of inter-group conflict, whether Nazis and Jews, Serbs and Croats, Japanese and Chinese, Hutus and Tutsis, Indians and Pakistanis, Caucasian Americans and North American Indians, Caucasian Australians and Aboriginal Australians, Palestinians and Israelis, or Irish Protestants and Catholics, one group has characterized the other as cargo, parasites,

---

24(3), 771-779; Fiske, S. (2009). From dehumanization and objectification to rehumanization. *Annals of the New York Academy of Sciences, 1167*, 31-34; Goff, P., Eberhardt, J., Williams, M.J., & Jackson, M.C. (2008). Not yet human: Implicit knowledge, historical dehumanization, and contemporary consequences. *Journal of Personality and Social Psychology, 94*(2), 292-306; Haslam, N. (2006). Dehumanization: An integrative review. *Personality and Social Psychology Review 10*(3), 252-264; Haslam, N., Kashima, Y., Loughnan, S., Shi, J., & Suitner, C. (2008). Subhuman, inhuman, and superhuman: contrasting humans with nonhumans in three cultures. *Social Cognition, 26*(2), 248-258; Lammers, J, & Stapel, D. (2010). Power increases dehumanization. *Group Processes & Intergroup Relations, 14*(1), 113-126; Moller, A.C, & Deci, E.L. (2009). Interpersonal control, dehumanization, and violence: A self-determination theory perspective. *Group Processes & Intergroup Relations, 13*(1), 41-53.

viruses, cockroaches, lice or vermin, or some mixture of these assassinations of humanity. The process is universal. The specific details of how other humans are stripped of their humanity is, however, up to the commanders of dehumanization. The second point is that the process of dehumanization is functional and strategic, certainly not an error of perception. When we dehumanize, whether unconsciously or consciously, it serves the specific function of removing moral constraints. Sometimes these constraints lift for beneficial ends as when a doctor thinks of surgery in cold, mechanical ways in order to push away the emotions of cutting into human flesh. Sometimes they lift for toxic ends as when a power hungry group demonizes a subordinate group to justify cutting off their flesh. With these two points in mind, consider a series of studies focusing on how whites have dehumanized blacks.

The social psychologist Jennifer Eberhardt asked whether American citizens unconsciously associate Black people with imagery of apes, using the disturbing history of this association as her jumping off point, as well as the many comments made by people to this day. For example, in December of 2011, Mirlin Toomer, a former black Defense Department worker, sought damages against a supervisor who hung a stuffed ape on her desk. Then there is the case of Chicago bartender Jessica Elizabeth who, on April 3, 2012, wrote several horrific comments about blacks on her Facebook page, including: "They are really apes and must not be fully developed."[35] Based on these comments, her boss fired her and apologized to the people of Chicago.

Eberhardt was interested in the possibility that if people carry this association around in their head, they may do so unconsciously, despite explicit avowals by some that they are not at all racist. And if they carry this association around unconsciously, how does it impact their judgments and actions?

In one experiment with both Caucasian and non-Caucasian subjects, Eberhardt used a technique called subliminal priming. Subliminal priming involves rapidly presenting pictures or sounds under the radar of a subject's awareness and then immediately presenting material that they are consciously aware of. If the two experiences are similar, the unconscious version will affect subjects' perception of the conscious one. For example, if you first prime people by flashing a picture of a woman's face, subjects will subsequently respond faster to faces of women than to faces of men. Thus, despite the fact that subjects are unaware of the prime, it affects their judgments. Eberhardt first primed subjects with photographed faces of Caucasian or Black people or an unrecognizable non-face. They then watched a short movie that started off with an unrecognizable object that looked like it was covered by dense snow. As the movie progressed, the snow lifted, making it easier to

---

[35] http://312diningdiva.blogspot.com/2012/03/proof-bartenders-racist-rant-put-on.html#more

recognize the object as a line drawing of either a duck, dolphin, alligator, squirrel or ape. Subjects stopped the movie as soon as they recognized the animal.

Compared with Caucasian faces and non-faces, priming with Black faces caused subjects to stop the movie much *sooner* for apes, but not for any other animal. Compared with non-faces, priming with Caucasian faces caused subjects to stop the movie much *later* for apes, but not for any other animal. This suggests that seeing Black faces made it easier to identify apes, whereas seeing Caucasian faces made it harder to identify apes. Critically, there was no comparable effects for any other animals. Caucasian and non-Caucasian subjects showed the same pattern of response, and so too did individuals with and without strong, explicit racial attitudes. Although the similarity among Caucasian and non-Caucasian subjects is of interest, and suggests that the association is held even among those who were perhaps less strongly associated with this form of dehumanization, there were relatively few Black subjects in the study.

This first set of experiments suggests that among a racially heterogeneous group of educated Stanford undergraduates, individuals carry an unconscious association between Black people and apes, and thus, an unconsciously dehumanized representation of another human being. Given the animal form of this dehumanization, and Haslam's analysis of the different dimensions of dehumanization, we can infer that subjects in Eberhardt's experiments associated Blacks with less rationality, civility, and self-control.

These are remarkably disturbing findings. They can't be explained by some superficial similarity between human faces and animals because Eberhardt found the same results when she presented either line drawings or words of animals. Had Eberhardt used actual photographs of animals, subjects could have used similarity in skin color or nose shape — for example, seeing a black human face would prime seeing a black ape face because both share the color black in common. Line drawings and written words cut the legs out of this account. Eberhardt's results suggest that we are biased to associate apes with the socio-cultural, racial category "Black."

These findings reveal a deep seated, dehumanized representation that is readily triggered even in highly educated people. But perhaps they are less disturbing then we might imagine. Perhaps the take home message is that we are closet racists with antiquated theories of evolution or God's design. Outside of these artificial studies, we are well educated citizens who keep our isms tucked away, locked up in our unconscious. Unfortunately, the unsettling feelings that many will have to these studies are exacerbated by an additional set of results collected by Eberhardt, linking unconscious impressions to harmful actions. Caucasian male subjects watched a video of a policeman using force to subdue a suspect who was either Black or Caucasian. When primed with an ape drawing, but not that of a tiger, subjects were more likely to say that the policeman was *justified* in subduing the Black suspect than the Caucasian suspect. These results suggest that dehumanization is recruited, often

Marc D. Hauser

unconsciously, to justify actions that are otherwise morally prohibited. These results suggest that we are more than closet racists. We are out of the closet, armed for prejudice and dehumanization.

To unconsciously think that Blacks are more like apes than other racial groups is to strip them of characteristics that are uniquely human. As Haslam notes, when we dehumanize others in this particular way, we no longer see them as human, but as wild animals, virulent parasites, or immature children, all lacking in intelligence, etiquette, rationality, and moral wherewithal. This mode of dehumanization is ancient, reflected in the writings and paintings of European explorers who encountered indigenous cultures in Asia, Australia, and Africa. Dehumanizing others into objects is equally ancient, cross-cultural, unflattering and dangerous. In studies carried out by Haslam, subjects judged objectified men and women as less capable of suffering and less deserving of moral compassion and protection, reinforcing the age old attitude we once held toward slaves and that many hold today toward prostitutes. When people become property, they fall outside the circle of moral patients. They are moral zeroes. Moral zeroes play no role in our judgments or actions, leaving us guilt-free to do as we please.

What I have explained in this section is that the capacity for dehumanization is part of human nature, an ancient capacity seen across cultures and periods of time. How individuals or societies deploy this capacity is shaped by experience and recruited as both a conscious and unconscious process. Whether we perceive others as animals or objects, we have subtracted important features of humanity from them, often in the service of satisfying a particular desire or in order to justify our own or another's actions. This process requires a combination of different neural systems, including those specifically involved in categorization, imagery, visual perception, memory, and moral deliberation. Studies of the brain reveal the mechanics of these dehumanizing transformations and highlight, once again, our mind's seemingly limitless capacity to creatively combine different thoughts and emotions — a central topic of chapter 3.

Brains without borders

As I discussed earlier, dehumanization enables doctors to treat their patients — human or nonhuman animal — as mechanical devices that require repair. This allows for cool-headed, rational, and skillful surgeries, while fending off the humanizing emotions of compassion and empathy. This is beneficial. This is a transformation that enables doctors working in war-torn areas or regions afflicted with a disease outbreak to treat hundreds of suffering patients as if they were treating cars on an assembly line. Good doctors allow their compassion and empathy to return as their patients regain awareness. Bad doctors maintain their cool, detached manner, insensitive to the physical and psychological pain of their waking patients. Bad doctors continue to perceive their patients like cars on the assembly line. Really bad doctors see their patients like cars that were created for personal R&D.

The cognitive neuroscientist Jean Decety showed that when physicians look at video clips of people experiencing pain from a needle prick, regions of the brain involved in pain empathy are quiet relative to non-physicians[36]. For physicians, it's as if they were watching a needle prick a pillow, or any inanimate object without feelings. These doctors have dehumanized their patients. Though we don't know how much experience was necessary or sufficient to cause the physician's lack of pain empathy, or the extent to which physicians are physicians because they were born with less empathy, Decety's findings point to individual differences in our capacity to feel what others feel and the potential modulating role of experience. Similar modulating effects arise in the context of our perception of racial differences.

Recall from earlier sections that the circuitry for pain empathy is less active when we see someone from a different racial group experience pain. Two other, similarly designed experiments, not only support this finding but extend it to another area of the brain — the motor cortex, involved in the production and perception of actions — and to the importance of unconscious racial prejudice. Caucasian and Black subjects watched a video of a needle penetrating a human hand while sitting

---

[36] Decety, J., Yang, C., & Cheng, Y. (2010). Physicians downregulate their pain empathy response: An event-related brain potential study. *Neuroimage, 50*(4), 1676-1682.

in the scanner. Consistently, subjects showed weaker activation in the pain and motor areas when watching the needle penetrate the hand from another race. This lowering of pain empathy and motor response for the out-group was greatest for subjects who, based on a survey, had the highest unconscious racial biases. Reinforcing a point I made earlier, we unconsciously and automatically make racial discriminations that guide our compassion for other humans. When we perceive someone from another race, regardless of our specific experience with them, they command less of our compassion. They are as deserving of our empathy as a surgeon perceiving a patient, or as a cooperator perceiving a cheater. Individuals outside of our racial group are dehumanized, often outside of our conscious awareness.

There is one potential snag in my explanation of the studies of racial biases and pain perception. By definition, we look more like those from within our racial group than those outside it. Perhaps the bias is less about race and more about those that don't look like us. To explore this possibility, a follow up study placed Black and Caucasian subjects in a scanner and presented them with a video of a needle penetrating a *violet*-colored hand. Violet hands are not only different, but far more different than either black or white hands in terms of our experience of skin coloration. Nonetheless, the activation pattern in the brain matched the subject's own race. When we feel less compassion for someone of another race, it is because of racial biases, not because of superficial differences in appearance. Color is simply a cue that reminds us of our prejudice.

The fact that we feel less empathy for people in pain if they fall outside our inner sanctum suggests that we have dehumanized them, stripping away dimensions of experience that humanize those within the sanctum. These are the dimensions associated with emotion, and when taken away, cause us to perceive the other as an object. Since objects can't feel pain or joy, we can't share in their experience because they lack experience altogether. If that is the case, then when we perceive any human group that has been dehumanized in this particular way, there should be little to no activity in those areas of the brain associated with thinking, feeling, wanting, and believing. To explore this possibility, the social psychologist Susan Fiske placed subjects in a brain scanner and presented photographs of either extreme out-group members, such as the homeless and drug addicts, or photographs of other groups such as the elderly, middle-class Americans, and the rich. When viewing the extreme out-group, there was little activity in an area critically related to self-awareness and the process of thinking about others thoughts and emotions — the *medial prefrontal cortex*. There was, however, intense activity in the *insula*, a brain area that is recruited when we experience disgust. As Fiske concludes, when we dehumanize the other to an object, we have effectively switched off the caring and compassionate areas of the brain. We no longer see humanity. We no longer have to uphold our moral standards because objects don't deserve such respect.

Fiske's results highlight both the variety of ways in which we dehumanize other human beings and the toxicity of this process. Built out of our biology, modified by our culture, and triggered by environmental events, the process of dehumanization causes us to see others as having less feelings, which causes us to feel less bad when they suffer. In fact, we don't perceive them as suffering at all because they are not the kind of thing that can suffer. Recalling the work of Gray, Wegner and Haslam, these are things without the dimension of experience, much more like God, robots, and dead people than they are like living men, women, children, dogs, and even people in a vegetative state. This process sets up another that takes us back to Corporal Graner and the atrocities at Abu Ghraib: moral disengagement.

The moral checkout counter

In *Moral Minds*, an earlier book of mine, I synthesized a large body of research showing that our moral response to events in the world often occurs automatically, driven by unconscious processes. These unconscious processes consist of emotions and a grammar of principles. When these processes operate, they generate moral intuitions about forbidden and permissible actions. The combination of these unconscious moral intuitions and our conscious recognition of moral norms and laws generates a rich system of constraints on our

behavior. These constraints comprise our moral standards. Sometimes, however, our standards disintegrate
as we pursue, consciously or unconsciously, a process
of moral disengagement. Freed from the guiding power
of our moral compass, we engage in behaviors that
carry no moral valence at all or carry moral benefits
of heroism and protection. Killing human beings who
have been dehumanized to objects or virulent parasites
leaves us cold. Killing thousands of individuals who have
been painted as the enemy leaves us proud, perceived
by those on our side as heroes. The scientific evidence
on moral disengagement[37], which I discuss next, helps

---

[37] Moral disengagement: Bandura, A. (1990). Selective activation and disengagement of moral control. *Journal of Social
Issues, 46,* 27-46. Bandura, A., Barbaranelli, C., Caprara, G.,
& Pastorelli, C. (1996). Mechanisms of moral disengagement in
the exercise of moral agency. *Journal of Personality and Social
Psychology, 71,* 364–374; Bandura, A. (2002). Selective moral
disengagement in the exercise of moral agency. *Journal of
Moral Education, 31*(2), 101-119; Bandura, A. (2004). The role of
selective moral disengagement in terrorism and counterterror-
ism. *In: Understanding Terrorism: Psychosocial roots, causes,
and consequences,* In: F.M. Mogahaddam & A.J. Marsella
(Eds.), Washington, DC, APA Press, pp.121–150; Jackson, L.
E., & Gaertner, L. (2010). Mechanisms of moral disengagement
and their differential use by right-wing authoritarianism and
social dominance orientation in support of war. *Aggressive
Behavior,* 36(4): 238-250; Kelman, H.C. (1973). Violence without
moral restraint: reflections on the dehumanization of victims

us further understand why desire recruits denial in the service of decimating innocent victims.

The psychologist Albert Bandura has demonstrated over decades of research that moral disengagement can arise in a variety of ways, including morally justifying our actions, denying personal responsibility, conceiving of our actions as less egregious than other possible actions, and by dehumanizing others. Though there are different pathways, all humans have the capacity to morally disengage, all of us do it to some extent, do it more often when we are young than when we are old, and engage in it often if our biology predisposes us and

---

and victimizers. *Journal of Social Issues, 29*, 25-61; McAlister, A.L., Bandura, A., & Owen, S.V. (2006). Mechanisms of moral disengagement in support of military force: The impact of Sept. 11. *Journal of Social and Clinical Psychology, 25*(2), 141-165; Obermann, M.-L. (2011). Moral disengagement in self-reported and peer-nominated school bullying. *Aggressive Behavior, 37*, 133–144; Osofsky, M.J., Bandura, A., & Zimbardo, P.G. (2005). The role of moral disengagement in the execution process. *Law and Human Behavior, 29*(4), 371-393; Shu, L.L., Gino, F. & Bazerman, M. (2011). Dishonest deed, clear conscience: when cheating leads to moral disengagement and motivated forgetting. *Personality and Social Psychology Bulletin* 37(3) 330–349; Shulman, E. P., Cauffman, E., Piquero, A. R., & Fagan, J. (2011). Moral disengagement among serious juvenile offenders: A longitudinal study of the relations between morally disengaged attitudes and offending *Developmental Psychology* 47(6): 1619-1632.

our culture encourages it. But why do we check out from our moral responsibilities?

We are morally engaged when we perceive that we may help or harm a moral patient, or that we want to help and avoid harming others. Moral patients deserve our care and concern, and in many situations we have a desire to deliver these moral goods. When we deliver them, we feel good. When we don't, believe that we might not want to, or consider options that may harm others, we feel bad, guilty, and remorseful; or at least we should if our moral compass is properly working. The tension between our beliefs and desires on the one hand, and the possibility that we may violate our own moral standards on the other, causes a ringing feeling of discomfort, what psychologists call *cognitive dissonance*. In such situations, moral disengagement can rescue us from our conflict. Moral disengagement liberates us from the anticipated feelings of guilt and remorse, enabling us to do bad things. Moral disengagement also liberates us from feelings of guilt and remorse that typically follow when we do bad things. Moral disengagement thus facilitates immoral behavior and allows us to feel okay when we act immorally. Moral disengagement allows people to rationalize harm by transforming lethal motives into morally justified and even benevolent ones. Moral disengagement allows us to excuse ourselves from moral responsibility, either disregarding the harm imposed or convincing ourselves that it was justified, even obligatory. Corporal Graner wanted to humiliate the Iraqi detainees in Abu Ghraib, to destroy their spirit so that they wouldn't fight back.

Satisfying this desire is non-trivial as it stands in direct opposition to typical moral standards. Graner did what he believed was necessary to do his job: by stripping the detainees of their clothes and hiding their faces with underwear, he removed their identity and humanity, and lifted the problem outside of the moral sphere. As the pictures of a smiling Graner revealed, he felt neither guilt nor remorse, but pride. By morally disengaging, Graner felt justified. Moral disengagement provides a visa for causing harm. These claims are supported by considerable scientific evidence.

In several cross-cultural studies of school-aged children, results consistently show that those who are most morally disengaged are most likely to engage in various forms of aggression, including bullying and repeated criminal offenses. These same children are also least likely to engage in helpful behavior, revealing that moral disengagement dispenses with the typical process of self-censure and sanctioning that we carry around when we are morally engaged. In a study of American prison personnel involved in death penalty sentences, executioners were more morally disengaged than support staff or prison guards. Executioners were more likely to dehumanize the convicted prisoner and provide moral and economic justifications. Executioners also felt less guilt because they had developed a narrative to justify their actions, one that ascribed complete fault and responsibility to the victim. Support staff flipped in the opposite direction, fully involved with the weighty moral issues associated with ending someone's life.

Overall, then, these studies show that those closer to actively harming others are more likely to fully checkout from the moral arena.

Moral disengagement *enables* behaviors that are either immoral, illegal, or counter to deeply rooted prohibitions against harming others. But moral disengagement can also arise as a *consequence* of immoral behavior. When consumers want to purchase products produced by unethical means, such as those created by abusive child labor, they morally disengaged after the purchase, finding ways to justify their decision by thinking that the abuse was either not so bad or exaggerated by the press. In a series of experiments by the psychologist Lisa Shu and her colleagues, subjects either read about scenarios where they had the opportunity to cheat and then decided whether or not to do so, or played a word game where they could earn extra money by cheating. Consistently, those who cheated were more morally disengaged than those who did not, often considering their actions as reasonable and justified given the circumstances. Those who cheated and morally disengaged also selectively forgot the rules of the game, mental gymnastics that help alleviate the dissonance we normally feel when breaking rules. Equally interesting, when subjects read about or signed an honor code, they were far less likely to cheat, morally disengage, or selectively forget the rules of the game.

These studies reveal the plasticity of our moral behavior, especially its vulnerability to being pushed toward highly unethical acts. They reveal how exchanging our present moral standards for a new set enables us to justify our actions, including acts of extreme violence that would have been unimaginable under the old regime.

Moral disengagement facilitates behavior that runs counter to our moral standards and eases the emotional pangs we feel when we violate such standards. Sometimes, this process has beneficial consequences as when it empowers otherwise fearful soldiers to go to war for just causes. More often, moral disengagement enables toxic actions by empowering rogue leaders to carry out genocide under unjust causes. Moral disengagement is a process that allows us to hibernate from our moral responsibilities. It is a form of self-deception, a partner to dehumanization in the denial of reality. But self-deception, like deception of others, is not always harmful. In fact, it is often highly beneficial.

Angelic denial

In a nationally televised address in 2005, the Iranian President Mahmoud Ahmadinejad pronounced that the Jews had "created a myth in the name of the Holocaust and consider it above God, religion and the prophets."[38] Judge Daniel Schreber believed that his brain was softening and that he was turning into a woman in order to form a sexual union with God. During a doctor's

---

[38] Iranian President Mahmoud Ahmadinejad: http://www.washingtonpost.com/wp-dyn/content/article/2005/12/14/AR2005121402403.html

visit, a man reported that his pet poodle had been replaced by an impostor, masquerading *as if* he was the real deal. Judge Patrick Couwenberg stated under oath that he received the Purple Heart for military operations in Vietnam, and soon thereafter carried out covert missions in Southeast Asia and Africa as a CIA agent. The pilots of Air Florida flight 90 ignored signs from their dashboard indicating engine trouble and then proceeded to crash into a bridge, killing 74 of the 79 people on board. In 2008, while Hilary Clinton was running for President of the United States, she regaled admiring supporters with stories of her international experience, including her visit to Bosnia in 1996 where her plane was forced to land under sniper fire, followed by a rapid evacuation for cover. When I was a teenager, I often walked onto the tennis court thinking that I was John McEnroe, serving and volleying like the world's number one player.

Each case above tells the story of a person who acted as if the world was one way even though it wasn't. The Holocaust and its trail of atrocities were real, con-firmed by thousands of scarred survivors and the rela-tives who have heard their accounts — this includes my father. Judge Couwenberg was never in Vietnam, never earned a Purple Heart, and never had a connection with the CIA. There are no pet poodle impostors. Our brains don't soften, though they do deteriorate with age. When dashboard indicators suggest engine trouble, better to be safe than sorry when you are responsible for the lives of many people. Hilary Clinton landed in the exceptionally *safe* airport of Tuzla where she was *warmly* greeted by US and Bosnian officials. I am no McEnroe.

In each of these cases, there was a mismatch with reality. The person either harbored a false belief, but believed it was true, or was outright lying. In some cases, the mismatch was due to psychosis, some kind of delusion or malfunctioning of the brain. These people didn't know that their beliefs were false. In other cases, the mismatch resulted from an intentional lie or distortion, a process that is adaptive, designed to promote self-confidence and manipulate others. When I conjured up images of McEnroe, I momentarily deceived myself. I believed it helped my game. I never thought I was McEnroe. When Hilary Clinton misreported her trip to Bosnia, perhaps she misremembered or perhaps she distorted her memory to convince voters that she had what was necessary to run the country — toughness and international experience.

Some cases of self-deception are harmless and even beneficial, as in my illusion of tennis grandeur. Others are only mildly harmful, as in Clinton's distortion of her political experiences. And yet others are deeply harmful, as when leaders such as Ahmadinejad deny the suffering of millions. Self-deception thus traffics between two poles, from harmless to harmful.

Why does our mind play tricks on us, allowing us to believe things that are false? Why didn't we evolve a reality checking device that is vigilant 24/7? The answer here parallels the refrain carried throughout this book: like dehumanization, self-deception is Janus-faced, showing

Marc D. Hauser

both an adaptive and maladaptive side. Self-deception allows us to protect ourselves from the reality of a current predicament or loss. Self-deception allows us to brand ourselves in a more powerful way, providing a competitive edge in attracting mates and garnering other resources. As Mark Twain presciently noted, "When a person cannot deceive himself, the chances are against his being able to deceive other people."

The evolutionary biologist Robert Trivers was the first to identify the adaptive significance of self-deception and its connection to deception. As argued in his book *The Folly of Fools,* what appears completely irrational about self-deception evolved as a consequence of selection to deceive competitors. The most effective self-deceiver acts on auto-pilot, without any sense of his true motives. Here, I further discuss these ideas, focusing especially on how deception works to satisfy our desires in the face of moral opposition[39]. As with the dangers of dehumaniza-

---

[39] Lies and self-deception: Carrión, R.E., Keenan, J.P., & Sebanz, N. (2010). A truth that's told with bad intent: an ERP study of deception. *Cognition, 114*(1), 105-110; Dhami, M. K., Mandel, D. R., Loewenstein, G., & Ayton, P. (2006). Prisoners' positive illusions of their post-release success. *Law and Human Behavior, 30*(6), 631–647; Fast, N. J., Gruenfeld, D. H., Sivanathan, N., & Galinsky, A. D. (2009). Illusory control: A generative force behind power's far-reaching effects. *Psychological Science, 20*(4), 502–508; Gino, F., & Pisano, G. P. (2011). Why leaders don't learn from success. *Harvard Business Review, April,* 68–74; Greene, J. D, & Paxton, J.

tion and its role in denying reality, so too is self-deception a dangerous state of mind, allowing individuals to inflict great harm on innocent others while feeling aligned with the angels. This process begins with two rather benign states of mind — our developing sense of self and our capacity to create positive illusions.

M. (2009). Patterns of neural activity associated with honest and dishonest moral decisions. *Proceedings of the National Academy of Science (USA), 106*(30), 12506-12511; Johnson, D. D. P., & Fowler, J. H. (2011). The evolution of overconfidence. *Nature, 477*(7364), 317–320; Johnson, D. D. P., Weidmann, N. B., & Cederman, L.-E. (2011). Fortune favours the bold: An agent-based model reveals adaptive advantages of over-confidence in war. *PLoS ONE, 6*(6), e20851; Karim, A. A., Schneider, M., Lotze, M., Veit, R., Sauseng, P., Braun, C., & Birbaumer, N. (2010). The truth about lying: inhibition of the anterior prefrontal cortex improves deceptive behavior. *Cerebral Cortex, 20*(1), 205-213; Moore, D. A., & Healy, P. J. (2008). The trouble with overconfidence. *Psychological Review, 115*(2), 502–517; McKay, R.T, & Dennett, D.C. (2009). The evolution of misbelief. *Behavioral and Brain Sciences, 32*, 493-561; Otter, Z., & Egan, V. (2007). The evolutionary role of self-deceptive enhancement as a protective factor against antisocial cognitions. *Personality and Individual Differences, 43*, 2258-2269; Robbins, R. W., & Beer, J. S. (2004). Positive illusions about the self: short-term benefits and long-term costs. *Journal of Personality and Social Psychology, 80*(2), 340–352; Wrangham, R.W. (1999). Is military incompetence adaptive? *Evolution and Human Behavior, 20*, 3-17.

It is not until the age of about 18-24 months that we acquire the ability to recognize ourselves in a mirror. It is not until a couple of years later that we have an explicit understanding that our own beliefs can sometimes differ from others that we interact with. It is not until this time that we develop the capacity to actively deceive, along with a powerful suite of social emotions that enable us to feel embarrassed, envious, and remorseful. These feelings link up our sense of self with our sense of others. These are comparative feelings and beliefs, and they feed back to who we are, either building up our self-confidence or crushing it. When my youngest daughter Sofia was ten years old, she announced that she will one day go to Brown University, earn a lot of money, have five children, obtain a veterinarian degree, open a restaurant, and win Olympic gold in gymnastics or soccer. Sofia was not delusional, but brimming with uncalibrated confidence. Her confidence was uncalibrated in part because this is who she is, and in part because she had no sense of what it takes to get into Brown, become rich, take care of five kids, obtain a vet degree, open a restaurant, and win gold. My wife and I would have been horrid parents if we had popped her bubble. We would have been irresponsible parents if we didn't, over time, describe the challenges associated with each of these desirable goals and help her move toward them.

Developing a sense of self depends on at least two capacities: looking inwards at what we know and are capable of doing, and looking outwards at what others know and are capable of doing. When we look inwards, if we honestly open our eyes to the richness of our autobiography, we will recognize cases where we have succeeded and those in which we have failed. This history reveals our knowledge and ignorance, our strengths and weaknesses, and our capacity to exert control or meld to external forces. When we look outwards, again with an honest, panoramic perspective, we learn about those who know more or less than we do, about those we can outcompete and those we lose to in defeat, and about situations that undermine our capacity for self-control. Distortion enters these personal narratives when we either lack information or filter it in some way, consciously or unconsciously, often for personal benefit.

Positive illusions are biases that distort our sense of confidence, control, and invincibility. Positive illusions differ from delusions in that they are less fixed, and more amenable to change. Delusions are highly maladaptive, a signature of brain dysfunction and the source of great personal suffering. Positive illusions, in contrast, are often highly adaptive, generating the confidence necessary to take on great challenges and challengers, convincing an audience or a group of opponents that we are stronger, smarter, and sexier. Positive illusions are a form of self-deception with considerable evolutionary benefits, as well as toxic consequences.

The criminology scholar Mandeep Dhami examined positive illusions in criminals incarcerated in prisons within the United States and the United Kingdom. Because recidivism levels are high among convicted criminals, with 40-60% of offenders re-convicted after 1-3 years from release, it is important to understand risk factors. Some criminals may believe that their prior offense was just a one-off event or bad luck, and that they will never engage in crime again. They have learned their lesson and feel confident in their capacity to lead a crime-free life. In their own eyes, they are low risk. Based on a sample of over 500 prisoners from medium security prisons, only 30% felt that they would commit another crime, and most felt that they were much less likely to commit another crime than other prisoners. Thus, whether prisoners were evaluating their own chances of success or their success relative to others, they were living with a distorted narrative, confident that they wouldn't commit another crime.

Certain experiences can also enhance positive illusions by giving individuals an unrealistic sense of self-control, along with a distorted expectation that future outcomes are highly deterministic. For example, people who are wealthy, highly educated, part of a dominant group, or citizens within a society that value independence, are more likely to believe that they have control over the future and are more likely to express optimism and high self-esteem. These attitudes often lead to a boosted sense of control and an illusory sense of control over future outcomes. The psychologist Nathanael Fast ran a series of experiments to further explore the relationship between power and illusory control, specifically asking whether subjects endowed with power expect control over outcomes that are strictly due to chance or that are unrelated to the domain of power.

Across each study, whether subjects recalled a personal situation where they were in power or had to imagine being in power, they were more likely than those in a subordinate position to express confidence about the outcome of rolling a six-sided die, predicting the future of a company, and influencing the results of a national election. Power distorts our sense of personal control.

That power and winning distort is a tale that has been told and retold countless times in the annals of industry and warfare. As Francesca Gino and Gary Pisano note, the business world is bloated with cases where leaders and leading companies crash because they fail to examine the causes of success. They assume, for example, that their success is entirely due to their brilliance, control over the market, and the weakness of the competition, as opposed to a shot of good luck. So too goes the story of unexamined war victories, as supremely confident generals discount relevant information about their opponents, leading battalions on a death march — a conclusion supported by the in-depth analyses of Johnson, Trivers and Wrangham.

Our willingness to accept victories without question stands in direct contrast with our motivation to scrutinize failures, drilling down for explanations or causes. When we lose or fail in some way, the negative emotions accompanying this experience focus our attention on working out an explanation. When we win, we bask in the glory, often blind to the underlying causes. Winning leads to confidence of winning again, accompanied by surges in the hormone testosterone. Sometimes,

Marc D. Hauser

confidence turns to overconfidence, a positive illusion that can have disastrous consequences.

Dominic Johnson explored the link between overconfidence and testosterone in the context of a simulated war game. Each subject played the role of a leader in a country at war with another over diamond resources. The goal of the game was to accrue the highest level of resources or defeat the neighboring country. Though war games on a computer can not capture the full reality of war, the fact is that military specialists throughout the world use simulations to prepare combatants for some of the strategic and emotional problems they will confront.

Most subjects judged that they would outcompete their opponents, had the desire to do so, with men more competitive than women. Those who believed that they would whip their opponents actually had the worst records, suggesting that they were not only uncalibrated but that their distortion of reality led to costly outcomes. Those with the highest expectation of victory had the highest testosterone levels and were most likely to launch unprovoked attacks on their opponents. Given these outcomes, and the use of such games by the military, perhaps they should be used as a screening device, selecting only those who pass the reality check.

Whether in real life or in the simulated world of computer games, brimming overconfidence can lead to a distorted sense of risk and the odds of victory in war — or any competitive arena. Though this is a costly strategy, there are clear evolutionary benefits under conditions explained by Trivers and Johnson. Self-deception is favored when opponents have imperfect information about their strengths and weaknesses, and where the payoffs are high relative to the costs. Self-deception leads individuals to go for it, convincing themselves and others that the risks are low, the gains are great, and the standard social norms are no longer applicable. This is a dangerous form of denial, a process of moral disengagement that enables us to satisfy our desires by justifying horrific means and ends. Sadly, this process has left one of its most toxic stains on our most innocent victims — children — under the flag of moral virtue.

During the tenures of Popes John Paul II and Benedict XVI, some 4,000 priests sexually abused some 10,000 innocent children. This is unquestionably an underestimate. This is excessive harm. The priests who carried out these horrific acts didn't intend to harm their innocent victims. They acted to fulfill a desire, while denying in important ways that they would hurt the children. They were morally disengaged, despite the constant reminders that they were men of the church. Popes John Paul II and Benedict XVI, together with their cardinals and bishops, were well aware of these rampant cases of child rape. They could have acted, but didn't. They could have reported the priests to the police for a criminal act, or minimally, could have defrocked them. Instead, they either let them keep their posts or moved them to another clergy. They convinced themselves through the magic of positive illusions that they could handle the matter internally and that all would be well again. Their omissions are archetypal examples of the sin of sloth.

Màrc D. Hauser

By omission, they are responsible for excessive harm and should be held legally accountable. This process has only just begun as evidenced by the decision in October of 2011 to indict Bishop Robert Finn for failing to report a priest who took pornographic photographs of young girls. Though Finn was only charged with a misdemeanor, this case opened a legal floodgate, followed in June of 2012 with the indictment of Monsignor William Lynn, a former cardinal's aide. Lynn knew about charges of pedophilia against 35 priests, and sent this information to Cardinal Anthony Bevilacqua, the senior member of the Philadelphia Archdiocese. These priests were never removed from their clergies, despite repeated complaints. Cardinal Bevilacqua died before the trial ended, but Lynn was sentenced to 3-6 years in prison. As Judge Sarmina noted, her punitive decision was based on the Monsignor's concealment of "monsters in clerical garb who molested children." She then followed up, looking right into his eyes, and said "You knew full well what was right, Monsignor Lynn, but you chose wrong."[40]

Sentencing these clergy members for omissions is only the start. It is a legal decision that should empower the parents and children who have suffered to rise up and demand justice for allowing excessive harm to occur.

---

[40] Judge Sarmina's verdict: http://usnews.nbcnews.com/_news/2012/07/24/12928246-philadelphia-monsignor-william-lynn-gets-3-6-years-for-cover-up-in-catholic-priest-sex-abuse-scandal?lite

It is a decision that should spread all the way up the hierarchy to Pope Benedict, as he too was aware of all the problems and yet did nothing to stop it. He could have morally engaged, rising above his self-interest in protecting the church. His decision to say nothing at all should cause everyone to express outrage over the fact that allowing priests to rape innocent children perpetuates a cycle of pedophiles as those who have been abused are likely to abuse others. The leaders of the church have not only committed a crime of omission, but have helped perpetuate a culture of harm. Popes John Paul and Benedict are evil omitters, allowing excessive harm to the most innocent of victims — our children.

Self-deception and deception are not only part of normal brain function, but adaptive processes that can yield significant dividends in the struggle of life. Like other behaviors that are tied to survival, self-deception can run out of control, as individuals are tempted to tell bigger and bigger lies in order to obtain bigger and bigger resources. Individuals can come to believe that they have super powers, and thus, are invincible, god-like. These self-deceptive beliefs can lead to personal ruin, along with the destruction of innocent lives.

$E = D + D$

My goal thus far has been to extract the universal core of our capacity for evil, identifying the elements or ingredients that are shared across all cases and all time periods. This is a minimalist approach, one that recognizes the variety of forms of gratuitous cruelty as well as the causes that trigger them. Some may engage in acts of gratuitous cruelty because of brain damage or developmental disorders, whereas others, who are perfectly healthy, engage because they feel threatened or wish to crush a competitor. I believe, however, that all of this variation can be distilled to two ingredients that are both necessary and sufficient parts of the recipe for evil: desire and denial. Some environments discourage this combination, whereas other environments encourage it. The important point is that all of us are endowed with the capacity to express our desires, deny elements of reality, and combine these two psychological states to cause both morally justifiable ends as well as horrifically immoral and unjustifiable ends. As noted, when a doctor has the desire to help his patient, he often recruits denial (in the form of dehumanization) to fend off emotions that can get in the way of clear-headed, precise surgery, operations that entail cutting into human flesh. This is justifiable. When a doctor has the desire to help his group and carries out experimental surgeries on those outside of the group by denying their moral worth, he facilitates cutting into human flesh, but for unjustifiable ends: other humans are simply not tools to be manipulated for personal satisfaction and knowledge.

The idea that we are all endowed with the capacity for evil, and that this capacity requires the combination of desire and denial, in no way implies that the expression of this capacity is inevitable. Even in environments that promote a psychology of evil, some will resist.

Resistance comes from individual differences in creating both unsatisfied desires and unconstrained systems of denial. Conversely, others are heavily predisposed to engage in acts of cruelty because their biology tilts them toward high risk, high reward, and low empathy. These differences originate in our biology, and with the evolutionary history of our species. This is our story, our history.

Marc D. Hauser

Recommended books

Baron-Cohen, S. (2011). *The Science of Evil*. New York: Basic Books.

Christakis, N.A. & Fowler, J.H. (2011). *Connected*. San Francisco: Back Bay Books.

Hamburg, D. (2008). *Preventing Genocide*. Denver: Paradigm Publishers.

Johnson, D. P. (2004). *Overconfidence and War*. Cambridge: Harvard University Press.

Kelman, H.C., & Hamilton, V.L. (1989). *Crimes of Obedience: Toward a Social Psychology of Authority and Responsibility*. New Haven: Yale University Press.

Mikulincer, M., & Shaver, P.R. (Eds.). (2011). *The Social Psychology of Morality*. Washington, DC: American Psychological Association.

Staub, E. (2010). *Overcoming Evil*. New York: Oxford University Press.

Trivers, R. (2011). *The Folly of Fools: The Logic of Deceit and Self-Deception in Human Life*. New York: Basic Books.

part ii:
one history

three: kingdom of cruelty

Man is the cruel animal. He is alone in that distinction.
— Mark Twain

In 1791, the Shawnee Indians of Ohio massacred sixteen white children and adults as revenge for a prior crime on their people:

> "Twelve children, two young men and a young woman had been stripped and lashed to trees and beaten to death with limber hickory switches which still lay on the ground nearby... All of them, down to the youngest child - a girl of about five - had been scalped. Fires at their feet had destroyed the legs and lower bodies of all... the Indians had indeed recognized Jacob Greathouse and they had reserved a very special death for him and his wife... Greathouse and his wife had been tethered each to a different sapling with a loop running from neck to tree.

Their bellies had been opened just above the pubic hairs and a loose end of the entrails tied to the sapling. They had then either been dragged or prodded around and around so that their intestines had been pulled out of their bodies to wind around the trees as they walked, Mrs. Greathouse had apparently died before getting much more that half unwound, but Greathouse himself had stumbled along until not only his intestines but even his stomach had been pulled out and wound into the obscene mass on the tree. They had been scalped and burning coals stuffed into their body cavities before the Indians departed."[41]

This is a spectacular display of cruelty, but by no means an isolated case in our history, nor a particularly creative example. Looking ahead one and half centuries later, and on the opposite side of the globe, we find the equally innovative techniques used by the Japanese in the massacre of Chinese citizens at Nanking in 1937: individuals were drenched in acid and left to disintegrate, alive; men's testicles and eyes were ripped out before burning them alive; babies were put on skewers and then tossed into boiling water; fetuses were cut out of awake women and then presented to them inside a jar of preservative. To many, these actions seem unimaginable. And yet we

[41] Eckert, A.W. (2001). *The Frontiersmen*. Ashland, Kentucky: Jesse Stuart Foundation, p.356.

are faced with the reality that our minds have an ancient track record of imagining them, over and over again.

Paradoxically, there are also cases where we seem incapable of killing others, even when we are licensed to do so. Detailed historical accounts of warfare reveal that soldiers often fail to kill even when this is their mandate. Such records also reveal that soldiers often experience extreme fear before firing the first shot, defecate and urinate in their pants when they do, and suffer from severe post-traumatic stress disorder when a war ends. These accounts tell a different story about human nature: we often have great difficulty killing other human beings, including cases where there are reasonable justifications to do so. In this sense, we are like most other animals who resolve disputes over resources by means of non-lethal aggression.

What requires explanation is how we evolved the capacity for not only killing but extreme cruelty, despite the internal brakes that sometimes operate to stop us. This is an explanation that requires an understanding of our evolutionary history as a species, including the nature of violence in other species and the factors that led to our own unique brand of violence.

The idea I develop, echoing Mark Twain's prescient quote in the opening of this chapter, as well as some of the ideas presented by the psychologist Victor Nell, is that we are the only cruel animal in the kingdom of animals, and we got here as an incidental consequence of our uniquely engineered brain. Unlike any other animal, our brain freely combines and recombines

Marc D. Hauser

thoughts and emotions to create a virtually limitless
range of solutions to an ever-changing environment.
The combination of desire and denial is one example,
a mix that enabled us alone to kill infants and adults,
friends and foe, lovers and competitors, family members and strangers, and with a vastness and level of
maliciousness that is unprecedented in the history of
life on earth. But before modern humans evolved this
capacity, millions of other social species found ways to
satisfy their desires for valuable resources, most often
by means of non-lethal aggression. This is part of our
inheritance. Understanding this evolutionary history
helps explain why many of our aggressive instincts
parallel those seen in other animals. Understanding
this history also reveals how far we have departed
from our last common ancestor — a chimpanzee-like
animal that could kill other species, as well as infants
and adults from its own species, but with nothing like
the gratuitous cruelty seen in our own species.

*HARMING OTHERS, version 1.0: the design of
non-lethal behaviors*
All animals are motivated to secure resources that will
enable them to survive and reproduce. At the most basic
and universal level, this is what life is all about. Gaining
access to resources enables individuals to accrue more
resources, live longer, and produce more offspring.
The path to acquiring resources is complicated by two
facts of life that were central to Darwin's insights into the

process of evolution: resources are limited and individuals must compete with others from the same and different species for these resources. Competition often triggers aggression and aggression often results in harm to one or more individuals. But for the majority of animals, the harm is non-lethal. This section describes the behavioral strategies that are in play when animals fight for valuable resources and why death-by-fighting is a rare event.[42]

Consider life on Earth before human existence, say 10 million years ago. Our closest living relatives the chimpanzees and bonobos are living in the forests of Africa, and so too are dozens of other mammals, birds, reptiles, amphibians, fish, and insects. And of course there are animals populating every other continent and

---

[42] Fighting and communicating a message: Bradbury, J. & Vehrencamp, S. (1998) *Principles of Animal Communication.* Sunderland, MA: Sinauer Associates; Bulbulia, J., & Sosis, R. (2011). Signaling theory and the evolution of religious cooperation. *Religion, 41*(3), 363–388; Espmark, Y., Amundsen, T., & Rosenqvist,G. (2000). *Animal Signals.* Tapir Academic Press; Hauser, M.D. (1996) *The Evolution of Communication.* Cambridge: MIT Press; Krebs, J., & Dawkins, R. (1984). Animal signals: mind-reading and manipulation. In: *Behavioural Ecology,* pp. 380-402, Sinauer Press; Maynard Smith, J. & Harper, D. (2003) *Animal Signals.* Oxford: Oxford University Press; Searcy, W.A. & Nowicki, S. (2010). *The Evolution of Animal Communication.* Princeton: Princeton University Press; Zahavi, A. & Zahavi, A. (1999). *The Handicap Principle.* Oxford: Oxford University Press.

the seas that surround them. Among the social animals — those living in groups — the common form of
aggression is one-on-one, and the context is typically
competition over food, a place to rest, or access to a
mate. Sometimes the aggression is initiated as an attack
and sometimes it is in self-defense. Sometimes it is highly
ritualized and planned, and sometimes it is a reactive
free-for-all. Sometimes it occurs within the group and
sometimes between. Severe injuries arise, but deaths
are rare. The aim is to resolve a competitive dispute by
means of non-lethal aggression, and if possible, non-
physical contact. If someone dies it is because an injury
leaves them incapacitated or vulnerable to disease. It is
not because their opponent aimed to kill. The ubiquity
of non-lethal aggression points to a suite of constraints
that prevent animals from killing their opponents.

It all starts with two or more individuals perceiving
a desirable resource within reaching distance. What
launches a first move and subsequently guides the process to its completion with a winner and a loser? In some
species there are rules of thumb that deflate any aggressive instincts before they are launched, even though there
are clear competitive interests. For example, in territorial
lizards and birds, if an emigrating individual lands in an
area and sees or hears another individual vigorously
displaying — push-ups with colorfully flashing neck
sacs in lizards, vocal arias in birds — they move on. The
rule: territory owners win, no questions asked. Another
rule of thumb arises in species organized around either
permanent or breeding-only harems: one male and many

females. Here, the rule is: harem master wins access to females, virtually all of the time. Two classic cases are the well-studied hamadryas baboons of Ethiopia and the elephant seals of California. In both species, males are much larger than females, with elephant seals providing an extreme case — the harem *master* can be ten times bigger than the females he mates with, and substantially larger than most males. In hamadryas, no one challenges the male over access to the females in his harem. Competition arises in acquiring females into a harem, a process that starts early, with individual males recruiting juvenile females. In elephant seals, either one or a few males completely monopolize the mating among the often hundred or more females within the harem. These males rule. As evidenced by genetic fingerprinting, virtually all of the offspring are sired by the alpha male, with leftovers going to numbers two and three. No mating competition. Competition arises when the young turks try to wear down the harem master through repeated challenges over the season. Eventually, often over the course of several mating seasons, the harem master loses a fight and hangs up his gloves.

These rules evolved to minimize aggression. When aggression arises it is because a once dominant animal has gotten weaker, either due to age or injury; this opens the door to challenges from younger and healthier individuals.

Dominance hierarchies, briefly discussed in chapter 1, provide another set of rules or norms that guide competition and thus, put limits on aggression. In general, irrespective of the species, high ranking animals out-compete low ranking animals for access to resources. If the rank distance between two individuals within the hierarchy is large, the subordinate acts like a migrating lizard or bird landing in a resident's territory: no contest, no competition, no fighting, no harm done. If the spread is less, say two individuals who hold adjacent ranks within the hierarchy, then other factors enter into the calculation. This is where the problem gets interesting as these other factors determine the start and end of a contest, and thus the landscape of potential harm.

Novel insights into the dynamics of aggressive competition emerged in the late 1970s and early 1980s due to two fundamental developments within evolutionary biology. The first was due to the evolutionary biologist John Maynard Smith who recognized that for any competitive interaction, there are different strategies, each with different payoffs. Some strategies are more costly, but return greater benefits. Others are more conservative and less costly, but return smaller benefits. How well any given strategy does depends on its frequency in the population, and thus, on whether the particular strategy is dominant or rare. This is the logic of games, and game theory developed by economists. Maynard Smith's central insight was to see these games as evolving over long periods of time, locked into epic arms races with predators battling prey, hosts competing with parasites, and males challenging each other for access to females. For example, consider a baboon troop with 20 adult males. Imagine that one of the males decides to bare his canines, stand up on his two hind legs, and charge *whenever* anyone comes near him and he is eating. This male is displaying his intent to attack at the slightest provocation. One could imagine that this would be very effective, especially if he is the only one displaying in this way. But if this display pattern spreads, and all other males do the same thing, then this strategy fails as it no longer distinguishes among the 20 males in the troop. The key insight from evolutionary game theory is that the effectiveness of a strategy depends on how common it is within the population. Power comes, in part, from being not too common or predictable.

The second development involved signaling theory and a challenge to the traditional approach that considered animal signals as truthful messengers of information. On the traditional view, when a monkey bares his canines, he is signaling his motivation to attack. When a dog puts his tail between his legs, he is signaling his submissive status. When a bird gives an alarm call, she is telling others that a predator is nearby. When a human smiles, he is conveying his desire for friendship. The new signaling theory presented a challenge to this honest view of communication. Why, for example, wouldn't individuals lie, deceiving others into believing that they were really tough, meek, in danger, or friendly, only to take advantage of the situation and gain added resources. This is deception, Trivers-style. Why, for example, wouldn't a baboon who was actually afraid, put on a tough-guy show and scare off his opponents? Why wouldn't a dog who was actually tough, send a

Marc D. Hauser

submissive signal at the start of the interaction, cause his opponent to lower his guard and then attack? Why wouldn't a bird send an alarm in the absence of danger, knowing that others will run for cover and leave all the food behind – no competition? Why wouldn't a human send a seductive smile to lure in an innocent victim for robbery? This line of questioning, developed by the evolutionary biologists Richard Dawkins and John Krebs, led to a number of studies showing that animals are engaged in a much more complicated and dynamic dance when they compete. Signalers attempt to manipulate their audience and the audience attempts to read the true intentions of the signalers. The question now becomes: how much should signalers deviate from their actual capacity and what can the audience do to best assess the signaler's honesty?

Static properties of the animal — its height, weight, tail length, antler size — indicate its raw, unfakeable ability to fight, what biologists call *Resource Holding Potential* or *RHP*. A red deer with a large set of antlers has paid the costs of growth, and is thus, a serious opponent with considerable strength. A tall, heavy, long-tusked elephant bull has spent the time and energy to bulk up, and can throw his weight around in a fight. Added to an animal's RHP are dynamic properties, features that require energetic investment in the moment such as the loudness or duration of a vocalization, or the height of a jump display. These dynamic properties form the foundation of competitive interactions and the raw material for assessments. When a resource is up for grabs and no simple rule of thumb or RHP factor trumps, animals assess each others' displays, attempting to work out what is real and what is bluff. This assessment is critical as it helps mediate the odds of being harmed.

The evolutionary biologist Amotz Zahavi provided a simple, yet far-reaching explanation of how animals — humans included — can determine the honesty behind the message: look at the costliness of the message as one sign of credibility. Signals gain honesty if they are costly to produce, where cost is relative to current condition or health. If every red deer can roar as loudly as the next one, then roaring carries no weight. It carries no weight because in every naturally living population, there is variation in physical condition due to limited resources and competition. Since roaring requires energy, the power and duration of roaring should be linked to an individual's condition. Only those who can afford to roar like mad will do so. Cheating is not possible on this view of signaling, or if it occurs, it is easy to detect; a red deer roaring like mad, but with a dull coat and ribs showing, will be exhausted after one bout and fail to provide a repeat performance.

Several studies support Zahavi's insight, including work on insects, crabs, birds, and gazelles, as well as hunter-gatherers and religious institutions. Hunter-gatherers do it by showing off and sharing their large prey capture, as well as by passing through extremely costly rites of passage from genital mutilation to body scarring. Religions do it by requiring their members to engage in elaborate, time consuming rituals: in a study of 83 utopian communities that were in operation in the 19th century, those carrying out the most costly rituals survived the longest.

What this section reveals is that animals have evolved a wide variety of non-lethal strategies to compete. They use rules of thumb, and engage in assessment in order to minimize the costs of battle. What this means is that after most battles, there is either no harm done or it is minimal. This is version 1.0 of HARMING OTHERS. This version encompasses most of the behavioral routines that animals use when they fight for resources, and this includes the human animal. All of these routines are controlled by specialized hormones, brain circuits, thoughts and emotions. Some of these physiological mechanisms have remained largely unchanged over evolutionary time, whereas others have changed, guiding the thrill of victory and the agony of defeat, together with differences in the willingness to take risks. Some of these changes inched animals closer to lethal aggression, pushed some right into it, and others over the top.

HARMING OTHERS, version 1.0: raging physiology

In any competitive situation, whether it is animals working out a strategy for maximizing the odds of obtaining food or humans working out a strategy for maximizing

the odds of check mating an opponent's king, someone will walk away as the winner and someone as the
loser. Winning feels good and losing feels bad. Winning
fuels confidence, including the kind of overconfidence
noted in the last chapter. Losing lowers self-esteem.
Depending on the opponent, including what they look
like and whether they are familiar or unfamiliar, it is possible to gauge the likelihood of winning or losing in
advance. Depending on the individual's prior history of
wins and losses, and details of his or her personality,
some individuals will embrace the challenge of a high
risk-high payoff strategy whereas others will adopt a
low risk-low payoff strategy. Winning, losing, and taking risks are all influenced by differences in hormone
levels, neurochemicals, and patterns of brain activation.
Some of these differences are set early in life by the
individual's biology, some change over the course of a
year, some within a day, and some within the period of
a brief glance that allows an opponent to assess the
competition. These physiological processes regulate
an individual's motivation to fight or flee, as well as
the sense of reward and loss that accompanies winning and losing. They adaptively regulate the capacity
to harm, at least until they malfunction. Malfunctions,
whatever their cause, can convert healthy, defensive,
competitive, and justifiable aggression into excessive
violence and in our own species, unethical violence.
This teeter-tottering between normal adaptive aggression and abnormal malfunctioning aggression is at the

Marc D. Hauser

root of my explanation of evil. Here we begin to put this part of the story together[43], looking first at testosterone

[43] Raging physiology, the evidence: Archer, J. (2006). Testosterone and human aggression: an evaluation of the challenge hypothesis. *Neuroscience and Biobehavioral Reviews*, 30, 319–345; Bernhardt, P. C., Dabbs, J. M., Fielden, J. A., & Lutter, C. D. (1998). Testosterone changes during vicarious experiences of winning among fans at sporting events. *Physiology and Behavior*, 65, 59–62; Flinn, M. V., Ponzi, D., & Muehlenbein, M. P. (2012). Hormonal mechanisms for regulation of aggression in human coalitions. *Human Nature*, 23(1), 68–88; Edwards, D. A., & Kurlander, L. S. (2010). Women's intercollegiate volleyball and tennis: effects of warm-up, competition, and practice on saliva levels of cortisol and testosterone. *Hormones and Behavior*, 58, 606–613; Edwards, D. A., Wetzel, K., & Wyner, D. R. (2006). Intercollegiate soccer: Saliva cortisol and testosterone are elevated during competition, and testosterone is related to status and social connectedness with teammates. *Physiology and Behavior*, 87, 135–143; Gleason, E. D., Fuxjager, M. J., Oyegbile, T. O., & Marler, C. A. (2009). Testosterone release and social context: when it occurs and why. *Frontiers in Neuroendocrinology*, 30(4), 460–469; Korzan, W. J., Forster, G. L., Watt, M. J., & Summers, C. H. (2006). Dopaminergic activity modulation via aggression, status, and a visual social signal. *Behavioral Neuroscience*, 120(1), 93–102; Mazur, A., Booth, A., & Dabbs, J. M. (1992). Testosterone and chess competition. *Social Psychology Quarterly*, 55, 70–77; Mehta, P. H., Wuehrmann, E. A., & Josephs, R. A. (2009). When are low testosterone levels advantageous? The moderating role of individual versus

— an evolutionarily ancient hormone that can loosen inhibitions, motivate fighting, and generate feelings of satisfaction when a victory is in hand.

Testosterone is a critical hormone in all social animals, guiding aggressive, sexual and social behavior. Testosterone surges when males defend their territories and when they recruit sexually receptive females. Stronger surges occur when individuals are challenged by competitors who want their territory, food, mates,

---

intergroup competition. *Hormones and Behavior*, 56, 158–162; Peper, J. S., van den Heuvel, M. P., Mandl, R. C. W., Pol, H. E. H., & van Honk, J. (2011). Sex steroids and connectivity in the human brain: A review of neuroimaging studies. *Psychoneuroendocrinology*, 36(8), 1101–1113; Salvador, A., & Costa, R. (2009). Coping with competition: neuroendocrine responses and cognitive variables. *Neuroscience and Biobehavioral Reviews,* 33, 160–170.; Salvador, A., Simon, V., Suay, F., & Llorens, L. (1987). Testosterone and cortisol responses to competitive fighting in human males. *Aggressive Behavior*, 13, 9–13; Salvador, A., Suay, F., González-Bono, E., & Serrano, M. A. (2003). Anticipatory cortisol, testosterone and psychological responses to judo competition in young men. *Psychoneuroendocrinology*, 28, 365–375; Sapra, S., Beavin, L. E., & Zak, P. J. (2012). A combination of dopamine genes predicts success by professional Wall Street traders. *PLoS ONE, 7*(1), 1-7; van Honk, J., Harmon-Jones, E., Morgan, B. E., & Schutter, D. J. L. G. (2010). Socially explosive minds: The triple imbalance hypothesis of reactive aggression. *Journal of Personality, 78*(1), 67–94.

or position within a hierarchy. This suggests that testosterone motivates animals to fight within the arena of competition. Conversely, testosterone drops during parental care, friendly interactions between mates, and coalitions among males. This suggests that lower levels of testosterone motivate animals to create or maintain social bonds, sometimes in the service of attacking others.

The role that testosterone plays in guiding aggressive competition is nicely seen by comparing winners and losers. Testosterone surges after an individual wins a fight, and drops following a loss. These changes are highly adaptive as they motivate winners to keep defending their resources, and motivate losers to give up and minimize future costs. Across a wide variety of species, humans included, winners are two times more likely to win the next fight whereas losers are five times less likely to win the next fight. In our own species, among male and female athletes, in sports including soccer, tennis and judo, winners have higher testosterone levels than losers because of winning. This effect even holds in competitive interactions with no physical contact, such as chess, dominoes and stock trading, as well as in cases involving audience members as opposed to those directly involved in competition. In a study of day traders on the London Stock Exchange, those making the highest profits had the highest levels of testosterone, and soccer spectators watching their team win the World Cup had higher levels than spectators on the losing side.

The ebb and flow of testosterone is thus an essential part of the story of our evolved capacity for non-lethal aggression. In humans, and all other social animals that are both closely and distantly related to us, surges in testosterone loosen restrictions on fighting, motivate resource defense, boost self-confidence following victory, and increase the odds of fighting again. These are all adaptive responses to living in a world of limited resources, and thus, a world where competition is necessary to survive. But one consequence of this adaptive system is that it can enable selfish, antisocial behavior toward others, while feeling good about it. Testosterone thus binds violence to reward. This is a connection that, if unrestricted, can lead to an appetite for cruelty. The puzzle is why we are the only species that lifts this restriction over and over again; the rest of this chapter explores this puzzle.

Testosterone is not alone when it comes to guiding individuals toward aggressive competition or away from it. The hormone cortisol regulates the stress response in fish, reptiles, birds, and mammals, including all ages of human mammals. When fear kicks in due to aggressive challenges from a dominant individual or from the appearance of a predator, cortisol rises. When individuals confront uncertainty, as occurs when they enter a novel environment or confront a stranger, cortisol rises. When cortisol levels are high, individuals are more likely to avoid costly situations or experiences, such as getting into a fight with a stranger or engaging in a social interaction with someone whose status is ambiguous as friend or foe. When cortisol levels are low, individuals are more aggressive, more reward focused, and less sensitive to punishment. Testosterone and cortisol therefore play within the bodies of animals like two children sitting on opposite ends of a see-saw. When testosterone is up and cortisol is down, individuals are primed to harm others and take risks. When testosterone is down and cortisol is up, individuals are risk averse, less likely to harm and more likely to engage in friendly social behavior.

The hormones testosterone and cortisol are joined by brain chemistry that can either dampen or heighten an individual's aggressive reaction. Serotonin is one of the central chemical agents in this process, working on aggression through the brain systems that guide self-control. High serotonin levels are associated with behavioral inhibition, whereas low serotonin levels are associated with disinhibition or impulsivity. These changes in inhibitory control start with genetic differences between individuals (detailed in the next chapter), are affected over the course of development by particular experiences, and ebb and flow with daily experiences. The most elegant evidence for the inter-play between genes and experience comes from studies of genetically engineered mice who have had a serotonin gene altered. In these mice, those with higher levels of serotonin are much less likely to attack an intruder than those with lower levels of serotonin. Serotonin thus impacts the motivation to harm others by shifting individuals along a continuum from more impulsive to more patient.

Can the genetic differences that guide the levels of serotonin, and thus guide the levels of aggressiveness, also override an individual's history of winning or losing? In other words, given the evidence discussed above that winners are more likely to win again and losers are more likely to lose again, can the genetic differences that determine serotonin levels cause some winners to lose and some losers to win? To answer this question, the biologist Norbert Sachser and his colleagues compared levels of aggression in genetically engineered mice with either high or low levels of serotonin, and a history as either winners or losers. Results showed that history mattered most: winners were more likely to engage in fighting an intruder than losers, whether they were engineered to have higher or lower levels of serotonin. This suggests a hierarchy of effects, with the experience of winning a fight dominant to an individual's genetic constitution, at least in terms of genes involved in the expression of serotonin; this is a theme that we will revisit in chapter 4 as it helps explain individual differences in our capacity to harm others, or to refrain from it.

Dopamine, as discussed in the previous chapters, is linked to the experience of reward, guiding both the individual's sense of when it will occur while simultaneously motivating behavior that maximizes the odds of obtaining the goods. When animals reach their goals or expect to obtain them, including food, mating, or winning a fight, the brain delivers a surge of dopamine. In lizards, dominant animals have dark rings around their eyes whereas subordinates do not. The dark eye rings signal to subordinates "back off." When animals see opponents with experimentally erased eye rings, they not only show increases in aggressive behavior but increases in dopamine, confirming the connection between this neurochemical and the anticipation of a successful and rewarding experience. As I explained in discussing the psychology of desire, when human subjects take L-dopa, a drug that increases the amount of dopamine, they feel more elated about an event in the future. In a study of stock traders on Wall Street (as opposed to London traders discussed earlier), results showed that those individuals with gene variants that cause higher levels of dopamine had longer careers as traders. To have a long career, you have to be willing to take risks and anticipate that more often than not your risky trades will pay off. Having more dopamine facilitates this risky business.

The general implication of work in humans and other animals is that individuals with heightened dopamine levels, whether naturally occurring or experimentally induced, are more likely to anticipate a rewarding experience and thus more likely to engage in risky behavior. This is true whether the risk entails finding food, competing for a mate, or fighting an opponent. What this means for our understanding of evil is that those who engage in over the top acts of gratuitous cruelty may do so in part because heightened dopamine causes them to anticipate feeling really good about it, while negating any potential risk. The conclusion is not "dopamine causes evil" but rather, "dopamine heightens the anticipated rewards of evil." This conclusion is also not an excuse, as there are individuals with heightened levels of dopamine that do not engage in any malicious behavior. The appropriate conclusion is that certain biological changes can increase the odds of violent behavior.

Like the other physiological changes I have noted, serotonin and dopamine do not operate in isolation. Rather they both interact with each other and are directly regulated by testosterone. Thus, surges in testosterone result in lower levels of serotonin and higher levels of dopamine. This makes sense in the heat of aggressive competition: when serotonin levels are low, restrictions on aggression are lifted; when dopamine levels rise, the anticipation of reward increases as well, joining the additionally rewarding properties of testosterone. Mice will work a response lever to self-deliver testosterone, and humans become addicted to testosterone ingestion. If you inject testosterone into a mouse while it is moving about, the location associated with the injection becomes tagged as a favorite spot in the landscape, a place to revisit. Drug abusers and gamblers, two personality profiles associated with heightened experience of reward and poor self-control, have elevated levels of testosterone and dopamine.

All of these changes in hormones and brain chemistry reveal why non-lethal aggression is rewarding. These same processes and others make lethal aggression rewarding as well, but only in a small subset of animals.

HARMING OTHERS, version 1.5: upgrade to lethal aggression

I noted in the Prologue that there are three situations in which animals kill others. Two are broadly distributed across the animal kingdom — predation and infanticide — and

Marc D. Hauser

one is extremely rare — adults killing other adults from the same species — adulticide. It is this rare form, as we will see, that is the most difficult to explain, but of greatest interest in terms of our own evolutionary history[44].

---

[44] Evolving lethal killers: Beckerman, S., Erickson, P., Yost, J., Regalado, J., Jaramillo, L., Sparks, C., Iromenga, M. (2009). Life histories, blood revenge, and reproductive success among the Waorani of Ecuador. *Proceedings of the National Academy of Sciences (USA), 106*(20), 8134–8139; Durrant, R. (2011). Collective violence: An evolutionary perspective. *Aggression and Violent Behavior, 16*(5), 428–436.; Ferguson, R. B. (2006). Tribal, "Ethnic," and Global. *In: The Psychology of Resolving Global Conflict: From War to Peace, Volume 1, Nature vs Nurture* (Ed. M. Fitzduff, C.E. Stout; Lawrence Erlbaum Ass.), pp. 1–15; Ferguson, R. B. (2011). Tribal warfare. In: The Encyclopedia of War (ed. G. Martel), Oxford: Blackwell Publishers, pp. 1–13; Gauthier, D., Chaudoir, N., & Forsyth, C. (2003). A sociological analysis of maternal infanticide in the united states, 1984-1996. *Deviant Behavior*, 24 (4), 393-404; Heinze, J., & Weber, M. (2010). Lethal sibling rivalry for nest inheritance among virgin ant queens. *Journal of Ethology, 29*(1), 197–201; Kelly, R. C. (2005). The evolution of lethal intergroup violence *Proceedings of the National Academy of Sciences (USA), 102*(43), 15294–15298; Maestripieri, D. (2011). Emotions, stress, and maternal motivation in primates. *American Journal of Primatology* 73: 516-529; Roscoe, P. (2007). Intelligence, coalitional killing, and the antecedents of war. *American Anthropologist, 109*(3), 485–495; Sussman, R.W. 1999. The myth of man the hunter, man the killer and

In virtually every taxonomic group of animals – insects, reptiles, amphibians, fish, birds, and mammals — there are predators and prey. Predators are not merely aggressive, but designed to kill prey species for the purpose of survival. When I say that predators have been designed, I mean that evolutionary processes have resulted in specialized brain areas, anatomical structures, and behavioral strategies that are highly

the evolution of human morality. *Zygon* **34:** 453–472; Tobena, A. (2011). Suicide attack martyrdoms: Temperament and mindset of altruistic warriors. *In: Pathological Altruism (ed. B. Oakley),* New York: Oxford University Press, pp. 1–20; Watts, D. P., Muller, M., Amsler, S. J., Mbabazi, G., & Mitani, J. C. (2006). Lethal intergroup aggression by chimpanzees in Kibale National Park, Uganda. *American Journal of Primatology, 68*(2), 161–180; Wrangham, R. (1999). Evolution of coalitionary killing. *Yearbook of Physical Anthropology* 42: 1-30; Wrangham, R.W. (2011). Chimpanzee violence is a serious topic: A response to Sussman and Marshak's Critique of Demonic Males: Apes and the Origins of Human Violence. (Vol. 1, pp. 29–50). *Global Nonkilling Working Papers*; Wrangham, R. W., & Glowacki, L. (2012). Intergroup aggression in chimpanzees and war in nomadic hunter-gatherers: evaluating the chimpanzee model *Human Nature* 23(1)pp. 5-29; Wrangham, R. W., & Wilson, M. L. (2006). Collective violence: Comparisons between youths and chimpanzees. *Annals of the New York Academy of Sciences, 1036*(1), 233–256; Wrangham, R.W., Wilson, M. L., & Muller, M. N. (2005). Comparative rates of violence in chimpanzees and humans. *Primates, 47*(1), 14–26.

adapted to the problem of prey capture. As noted by the psychologist Victor Nell and the neurobiologist Jan Panksepp[45], predatory killing is enabled by stealth-like attacks accompanied by brain circuitry that makes killing highly rewarding. These two features are important as they distinguish predatory killing from most other forms of aggression. In particular, when predators kill, it is not a reactive response to a victim, but rather, one that is planned, including the initiation of a hunt when the victim is out of sight. When predators seek out their prey, the dopamine systems of the brain are highly engaged. Recall from the last section, that dopamine is linked to reward, and especially the anticipation of reward. For predators, the anticipation of killing is the opposite of aversive — there are no internal brakes at all. Once a predator kills its prey, the brain's opioid system kicks in, a system that I discussed in chapter 1. The opioids deliver pleasure. For a predator, killing and consuming its prey is like smoking opium in humans — deeply rewarding.

The second, common form of killing is infanticide, carried out by both parents and recently immigrated males. When parents engage in infanticide, it is often a response to the lack of resources required to rear young. To enable this response requires overriding those

---

[45] Nell, V. (2006). Cruelty's rewards: the gratifications of perpetrators and spectators. *Behavioral and Brain Sciences*, *29*(3), 211–57; Panksepp, J. & Zellner, M. (2004) Towards a neurobiologically based unified theory of aggression. *International Review of Social Psychology* 17:37–61.

instincts that evolved to take care of young. The primatologist Dario Maestripieri suggests that this can occur under conditions of stress. Recall from the last section that heightened cortisol levels are often associated with stress, which typically results in animals backing down from an aggressive attack. But when cortisol levels rise and are sustained over long periods of time, heightened aggression often follows, turning parental care into parental violence. In these situations, parents may kill their infants. This is as true of the monkeys studied by Maestripieri as it is of humans living under conditions of extreme poverty and stress — in the developed world, the United States has the highest level of infanticide, and as one researcher suggests, this may be due to the fact that extreme poverty is in close proximity to extreme wealth, a contrast that makes it even more stressful for the poor. In contrast with parent-initiated infanticide, the situation for recently immigrated males is different. Here, there are no parenting instincts to get in the way, and a deep motivation to kill infants sired by other males. By killing these infants, not only does the newcomer obliterate the competition's fitness, but effectively reboots the female's sexual receptivity. This type of killing is more like predation.

Both predation and infanticide involve significant asymmetries in size or weaponry between attacker and victim, making the kill relatively cost-free. These two forms of lethal aggression stand in striking contrast with the third and far more infrequent form in which the attacker and victim are from the same species, both adults, and thus, comparable in size and weaponry. This kind of killing or *adulticide* only occurs in a small number of species, but the attacks are sufficiently frequent to count as part of the repertoire: ants, lions, wolves, chimpanzees and humans. The rarity of adulticide raises important questions about the evolutionary pressures that favored this upgrade to killing others, as well as the mechanisms that evolved to make it possible.

Battles among ant colonies are notorious for their organized attacks, designed to kill the enemy and minimize costs. What is distinctive about ant battles and the deaths that ensue is that they are coordinated, with success driven by group size. As the biologist Eldridge Adams has demonstrated, bigger groups are more likely to win, more likely to kill a higher number of their smaller opponents, and less likely to incur any fatalities. Despite the similarities between ant and human battles, two differences undermine the usefulness of this analogy for understanding the evolution of lethal aggression in humans: ants are only a very distant evolutionary cousin, subject to extremely different pressures of social life, and their cooperative efforts are largely among individuals who are virtual genetic clones. When humans go to battle, cooperation is largely among unrelated individuals who are complete strangers. Only chimpanzees commit adulticide in a somewhat similar fashion.

When chimpanzees go on a lethal raid, it is anything but impulsive. In fact, chimpanzee lethal attacks look more like predation than any other form of aggression.

Things start with a gathering of individuals, usually all males, followed by a stealth approach toward the victim. With the victim in view, the attackers move in quickly. Once they catch the victim, the attack is brutal, accompanied by frenzied screams and hoots, focused on body parts that are necessary for moving, communicating and reproducing. Though no one has measured changes in brain chemistry in chimpanzees, chances are that the same processes arise as when lions kill gazelles: dopamine most likely rises during the hunt in anticipation of reward, followed by opioid increases following the reward of the kill.

Chimpanzee attackers commonly have a numerical advantage over their victims, a ratio of at least three to one. This power imbalance reduces the costs of the attack by making it almost impossible for the victim to retaliate. Proof of this cost-benefit analysis comes from the fact that the attacking party rarely incurs injuries, whereas the victims rarely escape alive. In a well documented case from Jane Goodall's site in Gombe, Tanzania, one chimpanzee community gradually eliminated their competitors in the neighboring community without incurring any deaths or serious injuries. The benefit of these attacks is that the attacking community gains access to additional resources by weakening the competitive strength of their neighbors.

The suite of behaviors that accompanies coalitionary killing in chimpanzees has led several scientists, most notably Richard Wrangham, to argue that this form of lethal aggression in chimpanzees is an adaptation, with deep parallels to human warfare. Like human warfare, chimpanzee groups with numerical superiority typically outcompete those with fewer individuals. Like human warfare, chimpanzee attacks are deliberate and take time to unfold. Like human warfare, the benefit of lethal aggression in chimpanzees is that it weakens the power of a neighbor, thereby enabling those who win to gain access to valuable resources. Given the variation in access to resources, this is the kind of behavior that is subject to selection. On this view, we inherited the upgrade to version 1.5 lethal aggression from a chimpanzee-like ancestor.

The claim that our capacity for killing, especially in war, is an evolved adaptation, is anathema to many scholars in the humanities and social sciences. This response is triggered by the belief that biological explanations imply inevitability, and provide an excuse for the atrocities we create. For these scholars, war, and more generally, the high levels of killing observed among human populations, are recent, cultural concoctions, born out of human intelligence, the invention of projectile weapons, and high population density, to name a few. From this perspective, biology plays no meaningful role in our understanding of human violence. From this perspective, culture can either trigger war or turn it off completely, an idea developed by the science writer John Horgan in his 2011 book *The End of War*. From this perspective, killing in chimpanzees is nothing like killing in humans. These general attitudes echo the famous 1986 Seville Statement[46] on violence in which a group of distinguished scientists, including the ethologist Robert Hinde, the geneticist John Paul Scott, and the paleoanthropologist Richard Leakey, sidelined biology with the following five statements:

1. "It is scientifically incorrect to say that we have inherited a tendency to make war from our animal ancestors."

2. "It is scientifically incorrect to say that war or any other violent behaviour is genetically programmed into our human nature."

3. "It is scientifically incorrect to say that in the course of human evolution there has been a selection for aggressive behaviour more than for other kinds of behaviour."

4. "It is scientifically incorrect to say that humans have a 'violent brain'."

5. "It is scientifically incorrect to say that war is caused by 'instinct' or any single motivation."

The Seville Statement ended with the conclusion that "Just as 'wars begin in the minds of men', peace also begins in our minds. The same species who invented war is capable of inventing peace. The responsibility lies with each of us." In essence, understanding our

---

[46] Seville Statement on Violence, Spain, 1986: http://portal.unesco.org/education/en/ev.php-URL_ID=3247&URL_DO=DO_TOPIC&URL_SECTION=201.html

biology will not contribute to understanding violence and war because we *invented* war as well as peace, woven out of nurture's cloth and her infinite tapestry of cultural potential. These kinds of claims about the role of biology in human behavior are at best incoherent and at worst plain wrong. They are also dangerous because they imply a view of human nature that is infinitely plastic, unconstrained by both universal features of our biology, as well as individual differences that predispose some to extreme violence and others to extreme altruism. These points have been eloquently spelled out by the evolutionary psychologist Steven Pinker, most recently in his treatise on human violence *The Better Angels of Our Nature*.

What makes the Seville Statement and other claims like it incoherent is a set of false attributions to biologists about the role of biology in human behavior. It is important to get this right as it is directly relevant to our understanding of why we evolved the capacity for gratuitous cruelty. Statements 2-5 are accurate in that it *is* incorrect to say that war or violence are *genetically programmed*, subject to stronger selection *than other kinds of behaviour*, built into the brain as a *violent brain*, and based on *instinct* with a single, inevitable output. But serious biologist don't believe statements like these. The biologist Peter Marler famously spoke of singing in birds as an *instinct to learn*, while Pinker described our capacity to speak as a *language instinct*, capturing the pioneering insights of the linguist Noam Chomsky. A bird's instinct to learn does not mean that there is a one-to-one, inflexible mapping between genes or brain circuits and a specific type of song. All songbirds have the potential to acquire their species' song, and in some birds, such as mockingbirds and parrots, this capacity extends to acquiring the sounds of other animals and even sounds made by inanimate objects. But if there is no input at all, or if the bird is deafened, the output is deficient in structure, unrecognizable as a species-specific song. The same holds for the language instinct. Instincts are biological biases that enable certain capacities, while limiting the range of potential variation. Our biology allows us, but not any other species, to acquire language. This same biology sets up constraints, due in part to what our brains can keep in memory, what our ears can hear, and what our larynx can produce. Like songbirds, the specific content of what we say, whether in French or Vietnamese, is determined by where we grow up and who we listen to.

If there is any intelligible sense of *genetically programmed* or *instinct*, whether for violence, language, sex, or mathematics, it is that our biology provides us with the capacity to acquire these domains of knowledge and expression. This doesn't mean that violence, language, sex or mathematics are inevitable or fixed in their expression. There are thousands of languages, ways of having sex, and forms of mathematical expression. There are also thousands of ways of being violent, and equally, ways of counteracting such violence. But none of this takes away from the importance of biology, especially its role in constraining the form that these expressions take in different environmental settings. To think otherwise is just wrong.

The debate about version 1.5 of lethal aggression gains interest if we restrict the conversation to the similarities and differences between chimpanzee and human killing. Similarities often speak to our shared evolutionary history, and point to the pressures that favored this form of violence. Differences speak to both changes in our biology and the environments we confronted and created.

Those who argue that chimpanzee killing is nothing like human killing, especially in warfare, come from two different traditions. On the one hand are anthropologists such as Robert Sussman and Brian Ferguson who suggest that chimpanzee killing is infrequent, yields little benefit in terms of resources or competition, and is restricted to populations that are either artificially provisioned by humans or crowded in by us. They also suggest that the archaeological evidence for human warfare doesn't really begin until about 12,000 years ago. As Ferguson notes "To argue that war is a result of some sort of innate predisposition to wage it requires that war be practiced throughout our prehistoric past."[47] This date, so Ferguson continues, is too recent to invoke natural selection as a cause, and leaves unexplained

---

[47] Ferguson, R. B. (2006). Tribal, "Ethnic," and Global. *In: The Psychology of Resolving Global Conflict: From War to Peace, Volume 1, Nature vs Nurture* (Ed. M. Fitzduff, C.E. Stout; Lawrence Erlbaum Ass.), pp. 1–15; quote: p.g. 45.

why there is no earlier evidence of massive killing if our last common ancestors had this capacity.

These criticisms either fly in the face of contradictory evidence, have little to do with the original ideas, or conflict with the logic of evolutionary theory. Concerning chimpanzee killing, the evidence comes from multiple sites in East and West Africa, including sites with no provisioning and no crowding from humans. Further, analyses by Wrangham and his colleagues show that humans living as hunter-gatherers or subsistence farmers on the continents of Africa and South America, engage in coalitionary killing, using stealthy raids and imbalances of power to minimize the costs and maximize the benefits. Looking at thirty-two different small scale societies, calculations of the yearly median death rates were between 164-595 per 100,000. Looking at nine chimpanzee communities spanning five populations in Tanzania, Uganda, and Ivory Coast, the rates were between 69-287 per 100,000. Chimpanzees fall well within the range of human hunter-gatherers and subsistence farmers. This evidence not only suggests similarities between chimpanzees and human societies living under conditions most like our ancestors, but also provides a resounding rejection of the view that chimpanzee killing is infrequent and of trivial importance. If the rates of killing are comparable, then either they are trivial for both species or trivial for neither. Given that both chimpanzees and human hunter-gatherers live in small groups, killing even a few individuals can have a dramatic effect on their capacity to defend resources. Killing in both species is non-trivial.

Another similarity between chimpanzees and small scale human societies comes from analyses of two extreme warring cultures, the Waorani of New Zealand and the Yanomamo of Venezuela. Though violence accounts for between 40-55% of all deaths in these two groups, attackers appeared immune to injury, with no more than 5% dying in battle, and often no deaths at all. Chimpanzee attackers are likewise immune to injury, due in large part to the strategic use of imbalances of power.

The parallels between chimpanzees and humans living in small scale societies supports the idea that similar pressures favored the *capacity* for coalitionary killing in both species. But this does not imply that these species should always kill in this way, contrary to what Ferguson and others wish to conclude. To say that a behavior or bit of anatomy is an adaptation, is to say that it was designed to solve problems in a particular environment. If the environment changes, it may or may not work or may work less efficiently. It is also possible that the behavior persists even after it is no longer adaptive; this represents some of the best evidence for the genetic basis of a behavior. A no-killing period in the archaeological record, as Ferguson points to, is interesting, but in no way undermines the logic of an adaptation. Such periods are *predicted* on evolutionary grounds.

The economist Samuel Bowles and his colleagues provide a different argument against the proposed similarities between human and chimpanzee killing. Unlike

Ferguson and Sussman, Bowles is entirely sympathetic to biology but sees fundamental differences in the pattern of human killing and warfare. To explain these differences he invokes two important attributes of human societies that have only weak parallels in other species: large-scale cooperation with unrelated others from the same group, together with hatred, symbolic labeling, and the motivation to hurt all others outside the group. These two factors, what Bowles calls *parochial altruism*, may have paradoxically generated both greater levels of cooperation within groups and higher rates of warfare between groups. Those groups with the best cooperators acquired the greatest resources and experienced the fewest losses due to cheaters and other morally corrosive rogues. This power and inward-looking favoritism led to self-defensive emotions and behaviors, ultimately leading to lethal aggression toward those with different beliefs and values. Thus parochialism and altruism co-evolved, hand in hand, breeding prejudice as a result of group safety. We saw an earlier example of this in chapter 1 where I discussed the results of Fehr's studies of fairness in children: kids of all ages were more likely to share with familiar children from their school than strangers from another school. Thus, evolution handed us a capacity to be discriminating altruists, a capacity that breeds hatred for those not like us, which recruits violence toward those outside the inner sanctum.

Bowles' analysis is interesting and consistent with my explanation of how we evolved the capacity for gratuitous cruelty. Certain aspects of our capacity to harm others emerges as an incidental byproduct of other capacities, and once this dynamic emerges, the combination of these capacities can evolve and change. What Bowles' analysis misses, however, is the fact that parochial altruism could well be true, and so too could our shared capacity for killing with chimpanzees. As noted above, rates of killing among chimpanzees and several small scale societies are comparable, and so too are the costs and benefits to attackers and victims. This argues in favor of a shared history and a shared adaptation. It does not mean that all aspects of killing in humans are similar, or that the human mind froze in a chimpanzee state with regard to its capacity to kill. There are differences that reveal the signature of evolutionary change.

Unlike the lethal attacks by chimpanzees that are restricted to cases where groups attack lone victims, primarily from neighboring groups, we wreak havoc on a massive scale, with one on one, many against many, and one against many, including victims within and outside our core group. Unlike chimpanzees, even our young children have an appetite for violence that can be nurtured, as evidenced by the brutality of child soldiers around the globe. Unlike chimpanzees, individuals will sacrifice themselves for an entire group as evidenced most recently by suicide bombers. Our minds also generate ideological reasons to motivate violence at extraordinary scales — again, think of suicide bombers taking their lives for a God, as well as the reward of an idyllic afterlife. And when our minds break down, or when we are afflicted with particular disorders early in life, we are capable of experiencing bizarre appetites for violence, including the joy of eating the flesh of murdered victims. These novel and unanticipated ways of harming others are the result, at least in their origins, of new hardware that evolved only once in the history of this planet: a brain wired to combine and recombine thoughts and emotions.

I will explain this idea in three steps, starting with a description of our brain's special design. I then turn to a discussion of how our brain's design incidentally enabled our species alone to punish moral transgressors and feel good about it. I conclude with the incidental birth of version 2.0 of harming others, a form of lethal aggression that was both extreme and enjoyable — evilicious.

Creative combinations

To appreciate the significance of the human revolution in brain engineering, consider an example of a different form of intelligence that evolved for one, highly adaptive function: to brain wash and then destroy another organism for purely selfish reasons. This is the kind of example that caused Darwin to marvel over the nature of design, doubt the beneficence of God, reflect upon the cruelty of nature, and ponder the problem of evil.

In Brazil, there is a parasitoid wasp of the family *Braconidae* that lays its eggs inside a particular species of caterpillar. Once the larvae are fully developed, they hatch out of the caterpillar. But this isn't the end of the caterpillar's role as surrogate caretaker. Once the larvae hatch, they are treated to an unprecedented level of care from the caterpillar who, Gandhi-like, foregoes all eating and moving to protect its adopted young, including violent head-swings against any intruder.

The wasps' ability to hijack the caterpillar's brain is an exquisite example of evolutionary engineering, as well as selection for a specific function. Think of all the pieces that had to come together for the wasp to subvert the caterpillar into a parental slave: finding the right host, laying eggs inside of the host without getting caught, designing eggs that will develop inside the host and then emerge alive, and rewiring the host's brain so that it sticks around and defends the young of another species. This highly adaptive and myopic pattern of thinking runs throughout the animal kingdom and across different contexts: birds that feign injury to deter predators from their nest, but deceive in no other context; cheetah mothers who demonstrate to their cubs how to bring down prey, but never provide pedagogical instructions in other relevant domains of development; and chimpanzees that use stones to crack open palm nuts, but never use these tools for any other function, including as weapons against dangerous neighbors or potential predators. In each of these cases, the capacity evolved to solve a specific problem and does so spectacularly well. But there is little to no evidence that these capacities are used to solve any other problem.

Like other animals, we too are equipped with adaptive capacities that evolved to solve particular problems.

Unlike other animals, however, these same adaptive specializations are readily deployed to solve novel problems, often by combining with other capacities — a fluidity of intelligence that has been noted by others, especially the philosopher Daniel Dennett and the archaeologist Steven Mithen. Like wasps, we deceive, manipulate and parasitize others, often cruelly. But unlike wasps, this capacity is not restricted to one type of victim and one context. As long as the opportunity for personal gain is high relative to the potential cost, we are more than willing to deceive, manipulate, and parasitize lovers, competitors and family members. The same point applies to teaching and tool use, capacities that are virtually unconstrained in terms of context. Teaching occurs whenever a knowledgeable individual identifies an ignorant individual, whether the context is formal as in schools and organized sports, or more casual, as when parents inform their children about social norms, children share the latest apps with each other, and married couples educate each other about their likes and dislikes. When we invent a tool, it may start out with a clear and narrow function, but this in no way limits its use. A hammer is designed for tapping nails and removing them, but can also be used to break hard things, pry open stuck things, and hurt those we don't like if nothing better is available. In each of these cases, the range of functions is virtually limitless because of our brain's capacity to repeatedly combine thoughts and emotions from different domains of knowledge. The combinatorial options are vast. What changes in the brain enabled us, but no other species, to engage in this fluid thinking?

Dennett, Mithen and others who have addressed this question have often pointed to the critical role of language. They suggest that our capacity to create and string together words to express our thoughts provides the glue to connect up different ways of feeling, seeing, hearing, smelling, and remembering. That language provides this kind of glue is undebatable. That language enables fluid thinking also seems undeniable. But language itself is a system that is built up out of different and highly connected parts of the brain, and thus, different kinds of knowledge. When our brain composes a sentence, either before it is articulated or if it is simply formed as part of an internal monologue, the system that is required to form words must connect with the system that structures words into grammatically appropriate sequences. If we are motivated to communicate this expression, it is necessary to connect with yet another system of the brain, the one responsible for the library of gestures that turn linguistic thoughts into linguistic sounds or signs that others can understand. Language is therefore an example of our brain's capacity to connect up different brain systems, but it is only one of many that enabled our fluid thinking.

To understand what changed in the brain, it is useful to paint a few broad-stroke comparisons, and then narrow in on the details. We know, for example, that brain size changed dramatically over the course of our evolutionary history, ultimately reaching three times the size of a chimpanzee's brain with the appearance of the first modern humans, some 100-200,000 years ago. From the archaeological evidence, we can infer that some aspect of the internal workings of the brain — not simply size — must have changed at about the same time in order to explain the appearance of a new material culture of tools with multiple parts and functions, musical instruments, symbolically decorated burial grounds, and cave paintings. Before this period, the material culture of our ancestors was rather uncreative, with simple tools and no symbolism. The new material culture was heralded by a mind unlike any other animal. No other animal spontaneously creates symbols, though chimpanzees and bonobos can, to some extent, be trained to use those we invent. No other animal creates musical instruments or even uses their own voice for pure pleasure. No other animal buries its dead, no less memorializes them with decorations; ants drag dead members out of their colony area and deposit them in a heap, though this is driven by hygiene as opposed to ceremonial remembrance and respect. Only a species with the capacity to combine and recombine different evolved specializations of the brain could create these archaeological remains. This period in our evolutionary history marked the birth of our highly inter-connected, combinatorial brain.

The brain sciences help us see the fine details of this new species of mind. The comparative anatomists Ralph Holloway, James Rilling, and Kristina Aldridge have analyzed brain scans and skull casts of humans and all of the apes: chimpanzees, bonobos, gorillas, orangutans, and gibbons[48]. These species had a common ancestor approximately 15 million years ago. As distinctive species with a shared heritage, they represent a considerable diversity of mating systems, dietary preferences, use of tools, group size, life span, locomotion style, communication system, aggressiveness, and capacity for cooperation. Thus, gibbons are primarily monogamous, live in small family groups in the upper canopies, swinging and singing to defend their territories, never use or create tools, are omnivorous, restrict cooperation to within the family group, and show little aggression. Gorillas live in harem societies, knuckle walk on the ground, are folivores or leaf eaters, rarely use or make tools in the wild, show aggression primarily between harems, communicate with a diversity of sounds, and

---

[48] Ape brains: Aldridge, K. (2011). Patterns of differences in brain morphology in humans as compared to extant apes. *Journal of Human Evolution, 60*(1), 94-105; Bruner, E., & Holloway, R. L. (2010). A bivariate approach to the widening of the frontal lobes in the genus Homo. *Journal of Human Evolution, 58*, 138-146; Rilling, J. K., Barks, S., Parr, L. A., Preuss, T., Faber, T., Pagnoni, G., Votaw, J. (2007). A comparison of resting state brain activity in humans and chimpanzees. *Proceedings of the National Academy of Sciences (USA), 104*, 17146-17151; Rilling, J. K., Glasser, M., Preuss, T., Max, X., Zhao, T., Hu, X., & Behrens, T. (2008). The evolution of the arcuate fasciculus revealed with comparative DTI. *Nature Neuroscience, 11*, 426-428.

show limited cooperation even under captive conditions. Chimpanzees are promiscuous, omnivores who hunt for meat on the ground and in the tree tops, create a diversity of tools that are culturally distinctive between regions, communicate with a diversity of sounds, are lethal killers when they confront individuals from a neighboring community, and are cooperative especially in competitive situations. Despite this diversity, nonhuman ape brains are much more similar to each other than any one is to a human brain. What changed since we split off from our ape cousins is both the overall geometry of the brain in terms of the relative size of different components, as well as the connections both within and between these components. Some of the most spectacular changes evolved within the frontal and temporal lobes, as well as their connections to other areas of the brain involved in the control of emotion and stress. These circuits play a critical role in decision making, self control, short-term memory, social relationships, tool use, language, and yes, violence.

For detail, and further evidence of combinatorial thinking, we turn to imaging studies of healthy adults, maturational changes in children, and brain damage in patient populations[49]. Consider tool use, an exam-

---

[49] Combinatorially creative: Ebisch, S.J.H., Gallese, V., Willems, R.M., Mantini, D., Groen, W.B., Romani, G.L., Bekkering, H. (2010). Altered intrinsic functional connectivity of anterior and posterior insula regions in high-functioning participants with autism spectrum disorder. *Human Brain Mapping, 32*(7),

Marc D. Hauser

ple I briefly referred to above. Though a wide variety of nonhuman animals use tools, only humans create

1013-1028; Ioannides, A.A. (2007). Dynamic functional connectivity. *Current Opinion in Neurobiology, 17*(2), 161-170; Jackson, M.C., Morgan, H.M., Shapiro, K.L., Mohr, H. & Linden, D.E.J. (2010). Strategic resource allocation in the human brain supports cognitive coordination of object and spatial working memory. *Human Brain Mapping, 32*(8), 1330-1348; Meyer-Lindenberg, A. (2009). Neural connectivity as an intermediate phenotype: Brain networks under genetic control. *Human Brain Mapping, 30*(7), 1938-1946; Peeters, R., Simone, L., Nelissen, K., Fabbri-Destro, M., Vanduffel, W., Rizzolatti, G., & Orban, G.A. (2009). The representation of tool use in humans and monkeys: common and uniquely human features. *Journal of Neuroscience, 29*(37), 11523-11539; Preuss, T.M., Cáceres, M., Oldham, M.C., & Geschwind, D.H. (2004). Human brain evolution: insights from microarrays. *Nature Reviews Genetics, 5*(11), 850-860; Rozin, P. (1999) Pre-adaptation and the puzzles and properties of pleasure. In: *Well-being: The foundations of hedonic psychology.* Eds., D. Kahnemann, E. Diener & N. Schwarz, Russell Sage; Hauser, M.D. (2009). The possibility of impossible cultures. *Nature* 460: 190-196; Stevens, M.C., Pearlson, G.D., & Calhoun, V.D. (2009). Changes in the interaction of resting-state neural networks from adolescence to adulthood. *Human Brain Mapping, 30*(8), 2356-2366; van den Heuvel, M.P. Mandl, R.C.W., Kahn, R.S., & Hulshoff Pol, H.E. (2009). Functionally linked resting-state networks reflect the underlying structural connectivity architecture of the human brain. *Human Brain Mapping, 30*(10), 3127-3141.

tools that combine different materials, have multiple functioning parts, can be used for functions other than the one originally designed, and function in the context of survival, reproduction, and leisure. These are the signature properties of a combinatorial brain. When we look at the material culture of the most sophisticated animal tool user — the chimpanzee — we see tools that use a single material, have only one functional part, were designed for this function, and the function set is strictly limited to survival or reproduction. Something as simple as a pencil, beyond the chimpanzees' wildest imagination, consists of multiple materials — rubber, wood, lead, metal — was designed for writing but can be used for poking a hole or creating a drumbeat, and has two functional parts — lead for writing, rubber for erasing. When you put a human subject in a brain scanner and record activity during observations of tool use, what you see is an orchestrated coordination between different and connected brain regions. There is activity in regions carrying out spatial analyses, motor behavior, goal directed assessments, and object recognition, and much of this activity is fed forward to the frontal areas for storage in working memory as well as judgment and evaluation. And as we learned in the last chapter, when we see other humans as tools or mere objects, we lose the connection to those brain areas involved in representing others as social creatures with beliefs and desires. This suggests that our fluidity involves both creating and breaking connections between different brain areas.

Even resting human brains show signs of being wired for combinations. When you lie down in bed and close your eyes, but before you drift off to sleep, your brain — assuming *you* are a healthy adult — shows activity in a family of inter-connected brain regions called the *default network*. This is your brain at rest, but it is anything but at rest. Some of the most active areas involve those that are engaged when we evaluate social relationships, consider what others believe and desire, who they are, and how we might interact in the future. This same default network looks very different in children, as well as in the elderly: it is much less connected. Growing up is connecting up. Growing old is disconnecting. We gain the power of combinatorial thinking as we mature and lose it as we age.

If connection is key, then developmental disorders of the mind or physical insult to the brain should result in predictable loss. A brain imaging study of individuals with autism is revealing. Individuals with autism fall along a spectrum, from low to high functioning. Though this spectrum captures important differences, all inflicted with this developmental disorder have difficulty with social relationships because of difficulties understanding the beliefs, intentions and emotions of others. All of these capacities require a system that can integrate multiple sources of information. During brain scanning, individuals with autism show a striking reduction of activity in three connected areas of the brain: the *insula*, *somatosensory cortex* and *amygdala*. The insula is an area of the brain that is like a traffic cop, responsible for coordinating the flow of information in the brain, both where it's coming from and where it should go. The somatosensory cortex handles our body's sensations, including how aroused we are about our experiences. The amygdala plays a key role in generating feelings and placing value on our experiences. With the traffic cop asleep, and the body's sensations and feelings dormant, it is no wonder that those with autism lack empathy, can't understand what it means for someone to be in love, are befuddled by deception, and find the bombardment from our media-intense world truly overwhelming. The lack of connectivity among those with autism is proof that connectivity is necessary for fluid, creative, unconstrained thinking.

We don't know exactly when rich connectivity within the brain evolved. We also don't know why it evolved, in the sense of explaining the ecological and social pressures that would have favored this particular brain design. What we do know is that once we evolved our massively connected, combinatorial brain, we were liberated from the myopic and functionally narrow domains of thought and emotion that typify the animal kingdom. This change paved the way for remarkably creative ways of thinking, including a new pattern of harming others, with both beneficial and toxic consequences.

Incidental justice

Most forms of violence in the animal kingdom are, as noted, non-lethal, emerging in the context of resource competition: two individuals or two groups fighting over food, water, land or mates. In the much more uncommon situation in which individuals use lethal violence, resource competition nonetheless dominates. In both lethal and non-lethal cases, the violence often occurs because of an infraction — an individual violating a social norm. For example, when an intruder steps into a resident's territory or a subordinate attempts to mate within the alpha's arena, a fight often breaks out. These attacks look like punishment, designed to teach a lesson. And in many of these cases, they function in this way[50]. But like the other capacities described above for non-human animals, the capacity for punishment is myopic, restricted to the context of competition. In contrast, humans punish competitors who transgress and cooperators who cheat. When we punish, we harm those who deserve to be harmed, at least based on certain social and moral standards. And when we punish, we use both non-lethal and lethal means, including excessive, gratuitous cruelty. Once again, the capacity to harm in this particular way didn't evolve for punishment. Rather, punishment evolved as an incidental consequence of our brain's connected and combinatorial design. The evidence to support this idea comes from comparative work on other animals, together with observations of human brain scans[51].

---

[50] Punishment in animals: Clutton-Brock, T.H. & Parker, G.A. (1995). Punishment in animal societies. *Nature* 373: 209-216.
[51] Incidental justice: Boyd, R., Gintis, H., & Bowles, S. (2010). Coordinated punishment of defectors sustains cooperation and can proliferate when rare. *Science, 328*(5978), 617-620; Boyd, R., Gintis, H., Bowles, S., & Richerson, P.J. (2003).

The evolution of altruistic punishment. *Proceedings of the National Academy of Sciences (USA)100*, 3531-3535; Boyd, R., & Richerson, P.J. (1992). Punishment allows the evolution of cooperation (or anything else) in sizeable groups. *Ethology and Sociobiology, 113*, 171-195; Carlsmith, K. M., Wilson, T. D., & Gilbert, D. T. (2008). The paradoxical consequences of revenge. *Journal of Personality and Social Psychology, 95 (6)*, 1316-1324; de Quervain, D.J-F., Fischbacher, U, Treyer, V., Schellhammer, M., Schnyder, U., Buck, A., & Fehr, E. (2004). The neural basis of altruistic punishment. *Science, 305*, 1254-1258; Fehr, E., & Gachter, S. (2002). Altruistic punishment in humans. *Nature, 415*, 137-140; Gachter, S., Renner, E., & Sefton, M. (2008). The long-run benefits of punishment. *Science, 322*(5907), 1510; Henrich, J., & Boyd, R. (2001). Why people punish defectors weak conformist transmission can stabilize costly enforcement of norms in cooperative dilemmas. *Journal of Theoretical Biology, 208*(1), 79-89; Henrich, J., Ensminger, J., McElreath, R., Barr, A., Barrett, C., Bolyantz, A., Ziker, J. (2010). Markets, religion, community size, and the evolution of fairness and punishment. *Science, 327*(5972), 1480-1484; Henrich, J., McElreath, R., Barr, A., Ensminger, J., Barrett, C., Bolyantz, A., Ziker, J. (2006). Costly punishment across human societies. *Science, 312*, 1767-1770; Herrmann, B., Thoni, C., & Gachter, S. (2008). Antisocial punishment across societies. *Science, 319*(5868), 1362-1367; Rockenbach, B., & Milinski, M. (2006). The efficient interaction of indirect reciprocity and costly punishment. *Nature, 444*(7120), 718-723; Sigmund, K., Hauert, C., & Nowak, M. (2001). Reward and punishment. *Proceedings of the National Academy of Sciences (USA)*,

The first point to make is that the absence of punishment in cooperative situations is not because animals never cheat in this context. On the contrary, animals cheat in both cooperative and competitive contexts. For example, both lions and chimpanzees cooperate in group defense against dangerous neighbors. Some individuals cheat by lagging behind or failing to join in altogether. These cheaters never suffer any adverse consequences. This stands in contrast to the many cases in which individuals cheat in a competitive context and are physically attacked or shunned for it. Nonhuman animals thus have the capacity to recognize, harm, and potentially change rule breakers. And yet, these capacities and behaviors are not applied in the context of cooperative interactions. Myopic thinking rules.

We can begin to understand how and why we evolved our broad-brush capacity to punish by comparing competitive and cooperative situations. When an individual attacks a rule breaker in a competitive context, there is an immediate benefit: the attacker gains access to the challenged resource. In contrast, when an individual attacks a cheater in a cooperative context, the benefit is at best delayed, and often uncertain. If a lion attacked a laggard who failed to cooperate in group defense, the laggard might join in on the next opportunity or not. Punishment in a cooperative context thus involves more delayed and uncertain returns. Waiting,

98(19), 10757-10762; Trivers, R.L. (1971). The evolution of reciprocal altruism. *Quarterly Review of Biology, 46*, 35-57.

especially when the payoffs are uncertain, is hard for most animals, humans included.

Unique evolutionary changes in the human brain allowed us to exert much greater patience than any other animal, overriding the pull of the hedonistic now. These changes include brain areas involved in imagining the future, as well as suppressing the pull of current emotions and temptations. These changes didn't evolve *for* punishment, but they were readily deployed by this system of justice. We rely on creative strategies to place value on the future, including putting resources away so that we can't use them, and making verbal commitments that bind us to the future. These strategies, each emblematic of our combinatorial brain, help diminish the emotional pull of immediate temptations, while raising the attractiveness of future payoffs. This is a brain that can wait for the delayed benefits of punishment. This change was accompanied by another that made punishing a cheater feel good, immediately.

When we punish or get even with those who have acted badly, we feel a hedonic high, an experience captured by the metaphor "revenge is sweet but not fattening." As demonstrated by the economist Ernst Fehr, this is more than a metaphor. When we hand someone his just deserts, punishing someone for cheating, lying, or breaking a promise, our brain responds as if we handed ourselves just desserts, activating brain circuitry associated with reward, including those that are associated with the release of dopamine. In one study carried out by Fehr, two subjects played an exchange game for money.

One player — the donor — decided how much of the money to give to another; the other player had no say, simply receiving whatever was offered. A third player observed, out of view, the outcome of the exchange. In some cases, observers witnessed a fair division of the money and in other cases an unfair division in which the donor kept a disproportionate amount of the total. The observer then faced a difficult decision: leave the two players with their take-home earnings or use personal funds to take away money from the donor, returning it to the bank. Taking money away from the donor is a form of costly punishment. It is costly in two ways: the punisher loses money he could have kept, and the donor loses money that he unfairly kept in his previous exchange. It is also highly moralistic — the donor's action has no direct bearing on the observer, especially since the game is played anonymously. Thus, punishment, if carried out, is geared for the greater good, designed to teach those who play unfairly a lesson.

When donors kept a significantly larger portion of the original sum, observers punished, paying the costs. They also reported feeling good about taking down the cheapskates. Where was this honey hit to the brain coming from?

To answer this question, Fehr and his colleagues put people in a brain scanner and used a technique called Positron Emission Tomography or PET. This type of scanning provides an image of how much glucose is used up by different brain areas during a task. Higher glucose consumption occurs when there is higher activity in a brain region. When punishers decided to punish a selfish donor, glucose consumption increased in a region of the brain associated with reward: the *dorsal striatum*, a major part of the dopamine system. This region is also active when you eat ice cream, earn money, solve an unexpected problem, and develop cravings for cocaine. Punishers incurred a financial cost, but gained emotional elation and internal reward. Punishment, as a form of harm, feels good to members of our species.

This section started with a comparative puzzle: why do social animals physically punish in the context of competition for resources, but only humans punish cheaters in both competitive and cooperative situations? Why, more generally, is punishment in animals myopic whereas punishment in humans is unrestricted in its application? What the evidence suggests is that our combined capacity to delay gratification and feel good when we harm cheaters, enabled us to punish in any context. It provided us with the tools to not only repair a puncture in the system of norms, but to feel good about incurring costs with uncertain payoffs. This is yet another context in which our brains deliver rewards for harming others.

Once we evolved a brain that could punish in both competitive and cooperative contexts, it enabled us to solve another problem: large-scale cooperation among genetically unrelated individuals, including strangers. When other animals cooperate, they restrict their altruistic acts to close kin, or if they engage unrelated individuals, the number of recipients are small and familiar. By restricting costly altruistic acts to kin, individuals gain by helping their genetic fitness. By restricting their actions to a small number of unrelated but familiar individuals, they allow reputation to build, thereby reducing the odds of being cheated. When group size grows, and with it the number of unrelated individuals, many of whom are strangers, the odds of being cheated grows. Humans solved this problem, to some extent, by recruiting punishment. As Robert Trivers, Ernst Fehr, Samuel Bowles and others have noted, punishment enabled humans to harm cheaters, either physically or by exclusion from the group. Though punishment is costly, it benefits the punisher who gains in status and often, access to resources; it also benefits the individuals within the group by weeding out the cheaters who can undermine cooperative goals. One such cooperative goal is the unification of members of one group to wipe out another group. This is cooperation in the service of competition. It is a capacity that often recruits large numbers of unrelated strangers, linked by a core set of beliefs. It is a capacity that often enables excessive harm to those with different beliefs.

Only humans recruit the faculty of imagination to invent novel ways of inflicting excruciating pain before we kill those who have violated a social or moral norm. In cultures of honor, thieves and adulteresses are stoned, bludgeoned, and burned to death, often by close family members. In war, enemy captives are tortured to extract information and often for cruelty's sake alone. For both torture and the death penalty, we have invested considerable intellectual capital to invent devices that inflict the greatest amount of pain before terminating an individual's life. Though the Middle Ages in Europe are often associated with the horrors of famine and disease, the most horrific human disasters involved the torture devices created by the Christian judicial system. For virtually every body part, and especially, every orifice, there was an appropriate technology, including knee splitters, joint dislocaters, jaw, anus, and vaginal rippers, flesh stretchers, body impalers, head crushers, and breast mutilators. Humans designed these devices for one clear purpose: to bring excruciating pain and disfigurement to the victim.

Torture and execution were often public affairs, witnessed with raucous pleasure by hundreds. Such entertainment continues today, either live with real audiences, or broadcast over the internet for the benefit of millions of eager viewers. In 2010, the videotaped execution of Saddam Hussein received some 12 million hits on YouTube, with comparable numbers for other public executions taking place in Yemen, Iran, Iraq, and North Korea. As I discuss further ahead, the fact that we often use gratuitous cruelty in a public forum suggests that it plays a highly social function, one in which we attempt to impress with excess.

What we have learned in these last two sections is that the evolution of a highly connected and combinatorial brain enabled us to creatively approach problems that have stymied other animals. Relative to their myopia, we have a broad vision. This change in brain structure led to many incidental consequences. Of most direct relevance here is the observation that we evolved a capacity to feel good when we punish cheaters. This form of punishment, together with the cases of schadenfreude and revenge discussed in chapter 1, provide a suite of situations in which our brain rewards us for harming someone else or cheering while others carry out the violence. This intimate connection between feeling good and violence facilitated the evolution of HARMING OTHERS, version 2.0.

HARMING OTHERS, version 2.0: designed for Homo sapiens

In chapters 1 and 2, I described several cases where humans develop an appetite for harming others. Lust killers represent an extreme version, but other examples are not too far behind, including brain washed child soldiers and the brutal dictators that have placed their mark on every continent and in every epoch of history. These people often feel good when they harm others, and especially when they cause great pain along the way; sometimes they feel nothing at all. What I suggest is that this level of cruelty is only possible with the neural hardware that uniquely evolved in our species.

To see how version 2.0 works, let's return to some of the raging physiology that I discussed a few sections back. Recall that there are hormones like testosterone that surge when individuals win a competition, whether this involves deer banging antlers, humans banging fists, or chess masters banging minds. Along with testosterone's increase is an increase in dopamine, a decrease in cortisol and serotonin, and a decrease in frontal lobe activity and control. Within the environment of a combinatorial brain, this physiological ballet affects our sense of fairness, empathy, moral conscience, attitude toward retribution and justice, as well our willingness to engage in lethal aggression.

Brain imaging studies reveal that the prefrontal cortex plays an essential role in regulating our aggressive impulses. The prefrontal cortex has deep connections with other areas of the brain, including those that regulate our emotions, determine what others believe and intend, and place value on our experiences. When individuals respond aggressively to an unfair offer in a bargaining game, testosterone levels rise and activity decreases in a part of the prefrontal cortex associated with self-control. Testosterone's effectiveness in human aggression arises, therefore, through its capacity to turn down activity in brain areas critical for self-control. Further evidence for the importance of self-control, and these brain areas in particular, comes from studies of patients with damage to the prefrontal cortex. These patients not only show abnormal aggressive responses to direct insult, but also to such trivial matters as being offered a lowish offer in an experimental bargaining game. Putting these different studies together reveals that regions of the brain involved in the calculation of equity are closely tied to those that determine just deserts, which in turn are tied to those that enable our aggressive instincts. But unlike other animals, where aggressive responses to inequities are limited to non-lethal means, such constraints are often lifted in our case. To understand how these constraints are removed, we turn first to pathology and the case of psychopathology.

When we label individuals as "psychopaths" in casual conversation, we typically refer to people who are selfish, callous, and manipulative. These are all characteristics that fit with clinical reports as well as scientific analysis. When Hollywood portrays the psychopathic profile, these same traits appear along with a brutally violent character, capable of cold, gratuitous cruelty. The psychopath, in this sense, is the poster child for my definition of evil: an *individual who uses gratuitous cruelty to cause excessive harm to innocent victims*. Of most immediate relevance is the fact that the brain abnormalities that result in psychopathy — abnormalities that are due more to nature than nurture — are the same brain areas involved in the healthy condition. What we can learn from studies of psychopaths, then, is both the vulnerabilities of healthy individuals to carrying out acts of gratuitous cruelty, and ways in which the psychopathic mind is so very different.

Psychopathy, like many other clinical disorders including the case of autism discussed earlier, represents a spectrum with both extreme and mild forms, the latter represented by many healthy individuals at some point in their lives. Thus, many relatively normal people have acted selfishly for longish periods of time, often showing little regard for the feelings of others. Sometimes such callousness and self-absorption can lead to outbursts of violence, resulting in harm to innocent others. For the clinical psychopath, however, there is not only an extreme Narcissistic personality, but an inability to control impulses (associated with structural and functional abnormalities in the prefrontal cortex), a tendency for heightened aggression, a lack of remorse or empathy, and an inability to connect emotions up with moral norms. These processes may have their origin, at least in part, in genetic differences: psychopaths commonly have genetic variants that result in higher levels of dopamine within brain pathways that regulate self-control. As a result, psychopaths not only experience a higher anticipated reward in planning their actions, but have less self-control over what they do. These same genetic differences also underlie substance abuse, another case of poor self-control that ties us back to the arguments presented in chapter 1 on desire and addiction: when willpower to overcome temptation is weak, and the anticipation of rewarding experiences is high, excessive behavior commonly follows, aimed at satisfying desires. Together, this suite of genetic, neurobiological and psychological processes results in an individual that is morally disengaged. Psychopaths are not only freed from the emotions and thoughts that typically constrain our capacity for violence, but further freed from the neural brakes that stop us from hurting others.

The essential point to keep in mind is that psychopathy is a clinical disorder that presents a spectrum — like autism. Though clinicians often suggest that there are different types of psychopaths, each with their own unique clustering of psychological and behavioral problems, the diagnosis rests on the fact that the patterns of abnormal behaviors are consistent and predictable. In contrast, when we casually say that an individual is a psychopath, we are pointing to characteristics that are shared in common with the clinical type, but are more ephemeral in their expression. And the important point here is that a fleeting expression of narcissism, impulsivity, callousness, and moral disengagement can turn into a habit, one that causes great harm to others, especially those unlike us. As I discussed in chapter 2, we are equipped with a variety pack of mechanisms to take out the other, often without feeling any guilt at all. We dehumanize those unlike us or think of them as dangerous. Either way, we have paved a path of justification. This path is further deepened by yet one other corruption of an otherwise beneficent process, providing one final example of our brain's combinatorial creativity.

Among mammals, including humans, oxytocin is released in females during labor and breastfeeding, and in both males and females during social bonding and parenting. This has led many to think of oxytocin as the cuddle hormone. But when oxytocin surges within our distinctively human brain, it does something extra: it fuels hatred toward those unlike us. Experiments by the psychologist Carsten De Dreu suggests how.

Subjects first sprayed either oxytocin or a placebo up their noses, and then played a series of bargaining games, much like those described in the last section on punishment. When oxytocin shoots up the nose, it goes straight to the brain. When De Dreu analyzed his results, he found that those who sniffed oxytocin (relative to placebo) perceived in-group members (relative to out-group members) as more likeable, more human, more richly endowed with social emotions such as embarrassment, contempt, humiliation and admiration, more worthy of saving in an emergency, and more deserving of money in the bargaining game. In contrast, with oxytocin, they were more likely to use their own money to take away money from out-group members and were more likely to perceive them as non-human, bleached of the features of experience and agency that define our humanity. Oxytocin is thus two-faced, cuddling with its left profile and harming with its right. It is part of the artillery that facilitates our unique capacity to harm others.

This is a small sampling of the ways in which the human brain enables new forms of harm, including adulticide. We didn't invent lethal aggression. We share this capacity with a small group of animals that also kill other adults. But whereas these other species typically restrict their lethal attacks to situations in which one group greatly outnumbers another, typically targeting adults from a neighboring group, we evolved far beyond this pattern. We adopted the cost-benefit analysis that drives killing in other animals and applied it to killing in a virtually limitless space of homicidal opportunities. We kill when we outnumber our opponents or are outnumbered by them, attacking individuals within and outside our core group. We kill spouses, ex-lovers, stepchildren, those who believe in God and those who don't, the wealthy and the poor, kin and non-kin, and even ourselves if the cause is important enough. Virtually anything goes.

Our combinatorial brain opened a vista of harmful means, including the capacity to address a multitude of injustices. This is a capacity that may well have evolved for inherently good and justifiable situations, but has resulted in cases that are incidentally bad and unjustifiable. It is a capacity that evolved in response to growing pressures to balance inequities and take care of those who attempt to cheat society. It is a capacity that enabled us to engage in punishment in a broad range of contexts, righting wrongs and opening a new path to feeling good about harming others. It is also a capacity that enabled us to take advantage of the seemingly irrational, unpredictable and intimidating use of gratuitous cruelty to destroy our opponent's willingness to fight back. But like so many other processes that I have discussed in this book, our capacity to impress with excess is not restricted to violence, but deployed in the context of charitable giving and sex.

Impress with excess

"The[y] tried to bite him in the hindquarters, sides, and especially the testicles, while he in turn struggled. . . . [They] bit simultaneously at the loins, testicles, and anal region ... The mobility of the victim was much impaired by the four pursuers... Another two minutes later the [victim] had a large gash in the right loin, the testicles had been bitten off, and he stood as if in a state of shock. Occasionally he made some frantic movements and was able to struggle free... but then some member of the [group] would renew the attack. . . . Eight minutes after the [victim] had stopped running he went down and the [group] stood over him pulling out his insides. Another two minutes later, the [victim] died." [52]

This passage, though similar in many ways to the vignette about the Shawnee Indians at the start of this chapter, was written by the ethologist Hans Kruuk and describes a seemingly cruel attack by four hyenas on a wildebeest. But hyenas aren't cruel because they don't intend pain for pain's sake. All they intend, if they intend anything at all, is to bring down their prey with minimal cost. When they pull out a wildebeest's insides, it is not to inflict pain but to obtain nutritional gains. When the Shawnee pulled out Mr. and Mrs. Greathouse's insides, and tied them to a tree so that they could unravel themselves, their only goal was to inflict excruciating pain prior to death. This is excess and one explanation for it is Zahavi's theory of costly signaling that I discussed at the start of this chapter.

The basic idea is that high cost signals are honest indicators of the signaler's quality. They are honest because only those individuals in good enough condition can afford to expend time or energy displaying.

[52] Kruuk, H. (1972). *The Spotted Hyena*. Chicago: University of Chicago Press, p. 149.

Costly signals are also social, public displays, designed to influence others, with personal payoffs in terms of survival and mating. When a gazelle stots — a jumping motion that does little to advance its position — it does so when cheetahs are lurking. Only those gazelles in good physical condition stot[53]. When a gazelle stots, it is signaling to the cheetah "I am in such good condition that I am willing to handicap myself by bouncing up and down. Don't chase me. Chase one of the other gazelles who are running but not stotting." This signal is effective: cheetah are much *less* likely to hunt and kill stotting than running gazelles.

Stotting appears excessive, but in fact is an honest signal of power[54]. It is, to borrow a phrase from the

---

[53] Gazelles jumping, cheetah chasing: Caro, T.M. (1986). The functions of stotting in Thomson's gazelles: some tests of the predictions. *Animal Behaviour* 34: 663-684; FitzGibbon, C. D., & Fanshawe, J. H. (1988). Stotting in Thomson's gazelles: an honest signal of condition. *Behavioral Ecology and Sociobiology*, *23*(2), 69–74.

[54] Impress with excess: Boone, J. (1998). The evolution of magnanimity. *Human Nature*, *9*(1), 1–21; Chang, L., Lu, H. J., Li, H., & Li, T. (2011). The face that launched a thousand ships: the mating-warring association in men. *Personality and Social Psychology Bulletin*, *37*(7), 976–984; Griskevicius, V., Tybur, J. M., & Van den Bergh, B. (2010). Going green to be seen: status, reputation, and conspicuous conservation. *Journal of Personality and Social Psychology*, *98*(3), 392–404; Griskevicius, V., Sundie, J., & Miller, G. (2007).

economist Thorstein Veblen, "conspicuous consumption."   By flaunting their superior condition, throwing
away resources just because they can, these stotting
gazelles benefit in the long run, living longer and leaving more offspring who will inherit their qualities. These
ideas carry over into human behavior, and across a vast
range of contexts, unlike the myopia of the gazelle or
any other species.

Blatant benevolence and conspicuous consumption: When
romantic motives elicit costly displays. *Journal of Personality
and Social Psychology, 93*(1), 85–102; Griskevicius, V., Tybur,
J. M., Gangestad, S. W., Perea, E. F., Shapiro, J. R., & Kenrick,
D. T. (2009). Aggress to impress: hostility as an evolved con-
text-dependent strategy. *Journal of Personality and Social
Psychology, 96*(5), 980–994; Klauss, H.D., Centurion, J. &
Manuel, C. (2010). Bioarchaeology of human sacrifice: violence,
identity, and the evolution of ritual killing at Cerro Cerrillos,
Peru. *Antiquity*, 84: 1102-1122; Kulczycki, A. & Windle, S
(2011). Honor Killings in the Middle East and North Africa: A
systematic review of the literature. *Violence Against Women*
17: 1442-1464; Nelissen, R. M. A., & Meijers, M. H. C. (2011).
Social benefits of luxury brands as costly signals of wealth
and status. *Evolution and Human Behavior* 32(5): 343-355;
Pope, N. (2011). *Honor Killings in the Twenty-first Century.*
Hampshire: Palgrave Macmillan Press; Van Vugt, M., & Iredale,
W. (2012). Men behaving nicely: Public goods as peacock tails.
British Journal of Psychology, 1–11; Zahavi, A. (1975). Mate
selection: Selection for a handicap. *Journal of Theoretical
Biology, 53*, 205–214.

The anthropologist James Boone and evolutionary psychologist Geoffrey Miller suggest that human magnanimity evolved as did stotting, as an honest signal of wealth and power. It represents a desire to impress through wastage. Publicly handicap yourself in the short run to benefit your wealth and status in the long run. Hunter-gatherers who bring home large prey from a day of hunting don't make cryptic deposits for others, but make sure that their offerings are public. The Mayans didn't build the pyramids for personal enjoyment behind walled enclosures, but in the open, visible to potential enemies as displays of excessive power to create something really big and costly. Among the Kwakiutl Indians, those chiefs with the highest status publicly gave away or burned the largest quantities of their own possessions. Big tippers don't tip in private, but in the presence of those who can admire their lavish tips. Billionaires who give to charities, such as Donald Trump and Ted Turner, don't make anonymous contributions, but broadcast their altruism through social media outlets. Rappers such as JayZ, Puff Daddy, and 50 cent don't have absurdly lavish cribs with a six pack of sports cars because this is what they like, but because this is what they can show off on MTV. Flaunting, even at a substantial cost, provides a path to power. This is a club whose motto reads "Impress with excess."

To explore when people flaunt their status by throwing away resources that they could keep, the psychologists Mark Van Vugt and Wendy Iredale carried out a public goods game like the Carlsmith studies on revenge that I discussed in chapter 1. Like other bargaining games, this one also presented people with a conflict between feeding self-interest and helping others, but with an interesting social and sexual twist. Based on Darwin's theory of sexual selection, briefly discussed at the start of this chapter, Van Vugt and Iredale predicted that men would be more motivated to show off their capacity for conspicuous consumption in front of women because in the majority of species, males compete among each other for access to reproducing females, and females choose males on the basis of the resources they can provide. In the bargaining game, individuals either kept the money that the experimenter gave them, or contributed some or all of it to a public fund; their choice was expressed in front of an observer of the same or opposite sex. At the end of the game, the experimenter doubled the amount of money that players contributed and divided it equally among the players. On a purely selfish level, keeping all of your own money pays off, especially if others contribute to the public fund: you get your money plus a share of the fund.

Van Vugt and Iredale found that men gave more money to the public fund when a woman as opposed to a man sat next to them. Men also gave more money when the woman sitting next to them was attractive than when she was not, and contributed increasingly larger sums when they played in the company of other male competitors. None of these audience effects were observed for women. Van Vugt and Iredale conclude that a man's contribution to a public good is like the gazelle's stotting or the peacock's showy tail: an ostentatious, but honest display of power.

These studies, and many others, suggest that we often tout our wealth and power by showing off, burning through resources just because we can. Men do this more often than women, using their magnanimous gestures as a seductive recruiting technique. And many women are indeed impressed, in part because such costly displays are honest indicators of male status. Men are also more likely to use aggression as a means of both gaining status among women and defending their own personal honor. In cultures of honor, men not only fight other men to defend women, but often kill women if they refuse a pre-arranged marriage, commit adultery, or seek divorce. Not taking care of business in this way is a sign of weakness in such societies. Here, men aggress to impress. Here, men are morally justified by their culture's norms to kill women that have strayed from such norms. Here, atrocious acts are justified on moral grounds, treating women as pawns that can be manipulated by powerful men. Here, aggression is carried out in a public forum, involving bludgeoning with stones, the use of swords to behead the victim, and multiple gun shots fired at different parts of the body to prolong pain until death. Though honor cultures span different time periods and continents, many are driven by male desire for reproductive power and thus, control over female behavior. As I mentioned in the discussion of lethal killing in animals, and especially the case of infanticide, men are motivated to kill women who refuse marriage, seek divorce, or commit adultery because these actions point to a man who has lost control and thus, lost reproductive power over women. Killing such women allows men to regain control and seek women that are more likely to submit, a point made by numerous scholars. These historical analyses are further supported by experimental studies aimed at the idea that aggression not only defeats the competition, but elevates the status of winners by honestly displaying their power.

The psychologist Viktor Griskevius and his colleagues ran an experiment to better understand how a person's motivation to attain high status might lead to the use of aggressive displays that impress both same-sex competitors and opposite sex mating partners. Subjects read one of two scenarios designed to trigger feelings of competition or romance, and then had to decide what they would do in response to an acquaintance who carelessly spilled a drink all over them. Women responded to both scenarios by stating that they would exclude the careless acquaintance from any future social interactions. In contrast, men stated that they would physically strike the acquaintance, but their imagined willingness to act with such violence in the context of mating was only boosted when other men were present. Men's motivation to use aggression was thus fueled by the presence of other competitors, a mating arena in which the most impressive stand out. Aggression therefore serves two functions: it provides competitive muscle when resources are limited, and sends an honest signal to potential mates about the aggressor's capacity to incur costs. Rape genocide represents an extreme version of the use of aggression to impress, and a horrifying example of how evil actions function as a form of conspicuous consumption.

Whether we look back in time to the genocides in Bosnia-Herzegovina and Rwanda, or fast forward to the ongoing genocides in the Sudan and Congo, rape is a trademark of these conflicts. In every case, rape is not a byproduct of frustration or a desire to have sex, but a systematic and strategic technique used by the perpetrators to control the reproduction of women, either by killing or impregnating them. Rape genocides are also designed to strike fear and generate paralysis. Rapes are often committed in front of husbands and other family members, which crushes all sense of honor and power in the men, and strikes fear in all women and children who rightly imagine that they may be next. Often, those who are raped, and not killed, are kept in a relatively healthy condition until the child is born. This forces the victim into a series of moral nightmares: carrying to term an unwanted fetus, giving birth to an innocent child that looks like the rapist, wanting to end the child's life but faced with the strong cultural norms that prohibit abortion. Rape genocides exemplify the use of excessively cruel means to bring about excessive harms. They are acts of evil that function to intimidate innocent victims.

The historical observations and experimental studies I have reviewed provide a link between violence and power, including reproductive power or *fitness* in the language of evolutionary biology. These connections suggest that aggressive displays send important information about the aggressor's resources. Excessive acts of violence are just one step away in this process, and like conspicuous consumption for luxury items, not only reveal power, but an unlimited power and willingness to destroy. When millions are raped, slashed, burned, chopped up, and gored before dying, there are only two possible explanations: the responsible individuals are either clinically mad with no sense of moderation or healthy schemers who strategically use such means to destroy their victims. The mad men are the psychopaths and lust killers that I discussed earlier — individuals afflicted with either brain damage or developmental disorders. The schemers are like stotting gazelles, wanting to impress others of their awesome powers by performing high risk displays in front of an audience or by signaling that they are crazy and unpredictable, capable of unimaginable destruction. The schemers deploy proactive, premeditated, and cold violence — like predators. As the sociologist Wolfgang Sofsky noted in his commentary on the Nazi concentration camps "Individuals demonstrated commitment by acting, on their own initiative, with greater brutality than their orders called for. Thus excess did not spring from mechanical obedience. On the contrary; its matrix was a group structure where it was expected that members exceed the limits of normal violence."[55] Unfortunately for society — past, present and future — the Nazis are not an isolated case. As some of the

---

[55] Sofsky, W. (1993). *The Order of Terror.* Princeton, NJ: Princeton University Press, p.228

earliest archaeological evidence reveals, we have a long and globally dispersed record of cruelty. These horrific acts were carried out as both sacrifices to the gods and to showcase the perpetrators' incredible powers — an unbounded willingness to escalate. When Slobodan Milosevic, Radovan Karadic, and Ratko Mladic launched their ethnic cleansing initiative, they didn't just displace or kill Albanians and Croatians, they raped their women, gouged people's eyes, and repeatedly smashed their skulls before killing them off. Admiral Luis Maria Mendia, one of the leaders in Argentina's "Dirty war," convinced victims to board a plane under the pretext of a freedom flight, and then once in flight, threw them out of the plane, adding sheer terror to the brutality of their death. In the massacre of Nanking, the Japanese policy was explicitly cruel, with the goal of crushing the Chinese will to fight back. Japanese soldiers dumped victims into pits and buried them alive or poured gas over their bodies and then lit them on fire, all while smiling. Accounts such as these litter the pages of history, revealing the consistent effectiveness of this strategy to frighten and intimidate victims. They reveal that the desire to impress with excess is part of the human repertoire, a routine that we often call upon in cases of conflict.

Why oh why?

Why did we evolve the capacity for gratuitous cruelty? The answer begins, so I suggest, in a special property of the human brain. Some time after we diverged from a chimpanzee-like common ancestor the human brain was remodeled to allow creative new connections between previously unconnected circuits. Our newly connected brain enabled us to explore new problems using a combination of older, but nonetheless adaptive parts. Some of these novel explorations led to highly adaptive consequences, as when we developed the ability to self-deceive in the service of pumping ourselves up to do better in the context of competition; or when we invented new technologies to solve difficult environmental problems, such as using spears to capture prey at a distance; or, when we acquired the know-how to stockpile and enhance resources such as food, water and fertile land that are critical to individual survival and reproduction; or when we evolved the richly textured social emotions of jealousy, shame, guilt, elation, and empathy, feelings that motivate individuals to recognize the importance of others' well-being and interests and to correct prior wrongs; or, when we tapped into the rich connection between reward and aggression to punish cheaters trying to destabilize a cooperative society. But these same adaptive explorations also resulted in incidental costs that have destroyed the lives of innocent individuals. The capacity to deny others' moral worth enabled us to justify great harms, including self-sacrifice as living bombs designed to annihilate thousands of non-believers. The capacity to create advanced weaponry enabled us to kill at a distance, thereby avoiding the aversiveness of taking out those staring back. The capacity to stockpile resources led to the growth of greed, increasing disparities among members of society, the inspiration to steal, and heightened violence both to defend and to obtain. The capacity to experience social emotions such as jealousy led to blind rage and a driving engine of homicide, including cuckolded lovers who kill their spouses and stepparents who kill their stepchildren. The capacity to feel good about harming others enabled us to recruit this elixir in the service of causing excessive harm in any number of novel contexts, from ethnic cleansings to bizarre fetishes that include self-mutilation. And the list goes on. This is the yin and yang of a combinatorial brain. This is the natural history of evil, its ancestry and adaptive significance. It is a capacity that lives within all of us, but some of us are more likely than others to deploy it. This variation is also part of human nature, a critical component in the evolutionary process.

Recommended books

Buss, D. (2006). *The Murderer Next Door.* New York: Penguin.

Dennett, D. (1991). *Consciousness Explained.* New York: Back Bay Books.

Ellison, P. T., & Gray, P. (Eds.). (2009). *Endocrinology of Social Relationships.* Cambridge: Harvard University Press.

French, P. (2001). *The Virtues of Vengeance.* Kansas: University of Kansas Press.

Goldhagen, D.J. (2009). *Worse than War.* New York: Public Affairs.

Kekes, J. (2005) *The Roots of Evil.* Ithaca: Cornell University Press.

Kiernan, B. (2007). *Blood and Soil: A World History of Genocide and Extermination from Sparta to Darfur.* New Haven: Yale University Press.

McCullough, M.E. (2008). *Beyond Revenge.* New York: John Wiley & Sons.

Miller, G. (2009). *Spent: Sex, Evolution and Consumer Behavior.* New York: Viking Press.

Mithen, S. (1996). *Prehistory of the Mind.* London: Thames & Hudson.

Pinker, S. (2011) *The Better Angels of Our Nature.* New York: Viking Press.

Wilson, M. & Daly, M. (1988). *Homicide.* New York: Aldine Press.

Marc D. Hauser

Wrangham, R.W., & Peterson, D. (1996). *Demonic Males: Apes and the Origins of Human Violence.* Boston: Houghton-Mifflin.

Zahavi, A., & Zahavi, A. (1997). *The Handicap Principle: A Missing Piece of Darwin's Puzzle.* New York: Oxford University Press.

four: wicked in waiting

The wicked are estranged from the womb: they go astray as soon as they be born, speaking lies.

— Bible, Psalm 53:8

No one has a say over their genes or their parents, including the environment that is created on their behalf. For individuals raised in poverty, abused by parents or abandoned by them, there may come a time when it is possible to purge the past, rise above it, and lay down new tracks. Success in this endeavor depends upon biological potential and the environment's toxicity. Though every healthy human being acquires the same basic biological ingredients, individual differences in how our biology expresses itself can either provide immunity against toxic environments or deep vulnerabilities. The unlucky ones inherit genes that predispose to sensation-seeking and risk-taking, callous and unemotional attitudes toward others, weak self-control, and narcissistic leanings. With this lottery

Marc D. Hauser

ticket, it takes little to trigger a mind capable of gratuitous cruelty. And yet some resist.

Individual differences, rooted in our biology, are an important source of variation for natural selection. In fact, without heritable variation, selection has nothing to work with. As the evolutionary biologist Ernst Mayr noted, "He who does not understand the uniqueness of individuals is unable to understand the working of natural selection."[56] This chapter presents the scientific evidence on individual differences linked to the problem of evil. This evidence helps explain the source of individual differences, including its role in sculpting different personality profiles that either deviate greatly from societal norms or follow them to perfection. This evidence helps us understand that living among us, there are those who are poised to act in wicked ways, and others who will resist despite the relevant temptations.

What's normal?

Much of our fascination with evil stems from the distinct impression that evildoers are anomalies. Their actions are inhuman, unimaginable, rarely witnessed, and detrimental to our species' survival. This impression carries with it an assumption about what is expected or typical of our species, as well as what is possible. It assumes

---

[56] Mayr, E. (1982). *The Growth of Biological Thought: diversity, evolution, and inheritance.* Cambridge, MA, Harvard University Press, p.g., 46.

that those considered evildoers have thoughts, feelings, and desires that fall outside of the repertoire of an average human being. Their actions are unimaginable, so we think, because most human minds lack the capacity to imagine butchering human bodies, either for the fun of it or without any feelings at all. Like so many simple claims that go unchallenged, we should be puzzled by this one. We should ask: what's normal? The evolutionary history of each species' brain provides only a partial accounting of what the brain can do. To illustrate, consider the domesticated dog and its ancestor the wolf. Though dogs live with humans and are often raised by them, they never acquire a human language. In this sense, the domesticated dog is just like the wolf. But what dogs can do, with greater facility than any wild wolf, is understand a variety of human gestures such as pointing and the movement of our eyes. This capacity emerged following a period of human domestication. Wolves were not part of this selective regime. But — and this is the most interesting twist in the story — wolf puppies raised by human caretakers develop into adults that can read pointing and looking quite well. This tells us that even wolves evolved the potential to read human gestures, but only human environments favor this skill. This tells us that what animals express is not necessarily indicative of their potential. To uncover their potential, we must alter the environment or wait for such changes to happen naturally.

When we ask *What's normal?*, we are asking two questions: what is the evolved repertoire and what is the evolved capacity? The evolved repertoire tells us something about the relationship between a species' biology and the environments that have shaped their behavior. The evolved capacity tells us about a reservoir of behaviors that may only emerge in novel environments.

What's normal human behavior? The same distinctions apply to us as to dogs and wolves, with the extra complication that our species adds because of historical twists and turns orchestrated by legal, political, ethical, religious, and medical points of view. History presents us with hundreds of cases where an accepted normal mutated into an unaccepted abnormal, or where abnormal mutated into normal. In the United States, homosexuality was considered a mental disease before the 1970s, with its own entry in the *Diagnostic and Statistical Manual of Mental Disorders (DSMMD)*. It was considered abnormal, a deviation from our species evolved repertoire as a heterosexually reproducing species. None of the other primates exhibit homosexuality as a stable class of sexual orientation, nor do any other primates strictly engage in homosexual interactions; bonobos or pygmy chimpanzees have both homo- and heterosexual interactions, but there are no bonobos that only engage in homosexual interactions. Homosexuality was, therefore, considered unnatural. What is unnatural is bad, and in this case, morally bad. Two events turned this attitude around. The first was an underground movement of gay psychiatrists. The second was a discovery by Evelyn

Hooker who noted that the manual's classification entry was based entirely on clinical interviews of gay prisoners. Once Hooker and others carried out interviews with gay men lacking criminal records, the old classification scheme was effectively dead. Homosexuality was freed from its jail sentence as a mental disease — as abnormal — and transformed (at least in some parts of the world) into an accepted normal. Homosexuality is part of our evolved capacity.

When clinicians diagnose individuals with a mental disorder, they are making a statement about deviance, about what falls within and outside the range of normal or typical mental states. Unfortunately, there are no clear categories, no bright lines separating normal from abnormal or uncommon. As the distinguished psychologist William James noted, however, the best way to understand normal human functioning is through the mind of the abnormal or atypical. Let's follow this logic by returning to two clinical cases briefly mentioned in the last chapter: autism and psychopathy. The reason for looking closely at these two in particular is because both are characterized by compromised capacities for self-control and empathy — capacities that are deeply connected to the problem of evil.

Autism is a developmental disorder that is typified by difficulties understanding what others believe and feel, and to repetition of behavior. Some individuals appear entirely locked out of the world, rocking back and forth to their own internal rhythm. Others, diagnosed with a sub-class of the autistic spectrum known as Asperger's, are high functioning individuals such as Dr. Temple Grandin, who not only teaches college-level courses, but has done wonders as a spokesperson for autism, as well as the animal welfare movement. This range already tells us that autism is represented by a spectrum, once identified by purely behavioral measures, but joined today by genetic and neurobiological markers. The budding genetic evidence is particularly helpful for explaining the observed variation. For example, the MAOA gene, located on the X chromosome, is involved in the regulation of social behavior and has different forms that map to differences in brain activity and stress physiology[57]. The different forms correspond to the number of copies of the genetic material. This copy number is, in turn, partially responsible for the spectrum of autism observed, especially the degree of social dysfunction, including stress and aggression. Add this source of variation to the observation that the odds of autism are much higher in boys than girls, in children of parents with careers in math, engineering, and the physical sciences, and of fathers who reproduce late in life, and you have a package of factors that can tilt individuals toward one end or the other of this developmental disorder.

---

[57] The genetics of autism: Cohen, I. L., Liu, X., Lewis, M. E. S., Chudley, A., Forster-Gibson, C., Gonzalez, M., Jenkins, E. C. (2011). Autism severity is associated with child and maternal MAOA genotypes. *Clinical Genetics* 79(4), 355–362.

Once we admit to a spectrum and begin to pinpoint the factors that push individuals to stand on one end or the other, we must admit to admitting virtually everyone onto this spectrum. All of us, at some point in our life, have lacked sensitivity to the feelings and beliefs of others. All of us have been self-absorbed and locked out from the rest of the world. All of us have failed to express empathy and compassion to others, often repeating such failures over and over again. All of us have been a bit abnormal in this sense, due to differences in nature and nurture. All of us fall, on occasion, within the spectrum of autism as well as other disorders of the mind such as psychopathy.

Like autism, psychopathy is not one neat and tidy disorder, but a spectrum. As I mentioned in the last chapter, psychopaths are impulsive, narcissistic, and lacking in social emotions such as empathy, remorse, and guilt. These behaviorally defined characteristics are complimented by genetic and neurobiological markers, some pointing to risks in the pre-school years, and linked to the same MAOA gene noted earlier. The spectrum that defines psychopaths ranges from hyper-smart, calculating, and powerful politicians to low IQ, down-trodden, serial murderers. Everyone of us occasionally shows our psychopathic face: self-absorbed, impatient, manipulative, and uncaring. What is abnormal, then, is living with these characteristics, all the time. Clinically diagnosed psychopaths, like clinically diagnosed individuals with autism, have the characteristic traits as stable components of their personality. An honest clinician will tell you, however, that stability is difficult to define, as are the essential traits. An honest brain scientist will also tell you that, despite the observation that psychopaths have hyperactive dopamine brain circuits that may drive sensation seeking, along with smaller frontal lobe circuits that may minimize their sensitivity to punishment and the capacity for self-control, these differences are statistical. What "statistical" means is that if you were to stack up all of the brains with hyperactive dopamine circuits and smaller frontal lobes into one pile, most, but not all would be from psychopaths. You would also find psychopaths in the pile of brains showing normal dopamine activity and average-sized frontal lobes. These brain differences are interesting, but they are not yet like fingerprints, absolutely and uniquely distinctive and diagnostic of a disorder. Such honesty reveals the challenges we face in answering the seemingly simple question *What's normal?*

Lawyers, judges and juries face the same problem as clinicians, often relying upon documents such as the DSMMD to determine when someone has acted outside the range of normal behavior. But for legal cases, there are two relevant dimensions to the normalcy problem. The first concerns whether the supposed criminal was sane or insane. An insanity defense requires evidence of a disease or defect of the mind. It requires evidence that the individual lacked the capacity to appreciate the criminal nature of the act as well as the capacity to conform. This is the part that relies on the DSMMD, as well as clinicians who can testify based on their expertise.

The second concerns a more general understanding of what a prototypical or normal human would or could do in a given situation. The idea seems straightforward enough, but as I mentioned above, is only deceptively straightforward.

Crimes of passion provide a useful illustration of the challenges we face, especially with respect to understanding how harm is ignited in the face of moral norms against it. Highlighting the truism that love makes you crazy, the crime of passion defense is invoked for cases where, in the heat of the moment, an individual finds and kills his or her spouse in bed with a lover. The defining feature of a crime of passion is that it was not planned and most people faced with the same situation would act similarly, unable to control their emotions.

The crime of passion defense seems straightforward. Like autism and psychopathy, however, it too relies upon a diagnosis of what a prototypical or average person would do in the same situation. This diagnosis requires an understanding of two difficult mental states: planning and self-control. Planning involves imagining the future, time traveling to a new world, dreaming up what we might do and how we might feel. We plan in the short and long term, filling up our mental sticky notes with to-do lists. Self-control enters into planning because what we imagine for ourselves — what we desire — is often inappropriate or unethical because it harms others or ourselves. As noted in chapters 1 and 2, the capacity to keep desire in check relies on moral engagement. Moral engagement requires self-control. Moral disengagement requires denial in order to loosen the grip of self-control and enable desire to have its way.

When Lorena Bobbitt cut off her husband John Bobbitt's penis, she fulfilled her desire to harm another. She carried out this gruesome act despite the moral and legal sanctions against it. But she did not plan this act in advance and nor did it occur in the heat of the moment, triggered by finding her husband in bed with a lover. It followed in the wake of his repeated philandering, attempted rape and psychological abuse. As an act, it fell between the cracks of a long-term plan and a reflexive response — it was hatched on the night of the fatal attack, triggered by seeing a carving knife in the kitchen. Lorena either lost self-control for that fatal moment or she was in complete control, aware of what she was about to do and justified by her own moral convictions, believing that harming John was just deserts. John was most definitely not innocent. The jury delivered a "not guilty" decision, appealing to a crime of passion defense. This decision effectively excused Lorena's harmful act as normal and justified given the mitigating circumstances.

When we consider our sense of evil in the world, we must pause to consider our own biases and prejudices about what's normal. We must ask about the human potential, about our evolved capacities and our ability to behave in novel ways in novel environments. When we say that a person, group or nation is evil, we are saying something important about human nature, about our capability as a species. We are saying something important about the relationship between nature and nurture.

Evil eggs and corrosive coops

Statistics collected by the Federal Bureau of Investigation show that there has been a steady decline in violent crimes — murder, forcible rape, and aggravated assault — committed by 10-18 years old males from 2001-2010, and virtually no change in violent crime rate among females of the same age. The total number of violent crime arrests of youths in 2010 was 75,900, with 82% committed by males. Given calculations of criminal recidivism rates, somewhere between 40-60% of these youths will repeat the same crime, try something new, or escalate. Many of these individuals are on a path to becoming career criminals — a career that criminologists estimate costs society approximately $1.5 million per individual. How does a career of violent crime start? Are there early warning signs? How early? How much starts with the egg and how much with the coop in which it was raised?

Early scientific interests in this chicken and egg problem can be traced to the efforts of the Italian physician and psychologist Cesare Lombroso. In 1876, he published his magnum opus *The Criminal Man*. This was a serious, scholarly book aimed at understanding the biological origins of crime. Based on measurements of both anatomical and psychological characteristics, Lombroso concluded that criminals were born not made.

Their defining features were throwbacks to our evolutionary ancestors, dehumanized by biological defects. Modern man was civilized and elegant. Criminal man was barbaric, a savage with slanted forehead, jutting jaw, and excessively long arms. Criminal man was more ape-like than human-like — a claim that harkens back to the Eberhardt studies of dehumanization that I described in chapter 2. Because the cause of these differences was biological, Lombroso argued that a life of crime was inevitable. Change through rehabilitation was hopeless. To protect society, these natural born criminals had to be taken out of society, either locked up or executed. These ideas formed the basis of several eugenics' movements, with the aim of weeding out the undesirable, less than human elements of society, be they less intelligent, not white, and not from the right religious denomination.

Lombroso's theory of criminality was soon rejected as scholars from a variety of different disciplines unearthed its racial prejudices and shoddy methods, including a failure to include the many people with slanted foreheads, jutting jaws, and long arms who never committed crimes, and those with statuesque anatomy who did. This initiated a general skepticism and even fear of biological explanations, causing a swing in the opposite direction. Criminals were not born but made by corrupt societies. Humans are not born with biologically encoded scripts for behaving with malice or virtue. Rather, we are born with blank slates, waiting for society to inscribe its distinctive signature. So began a pendulous swing from nature to nurture explanations of human behavior. Though the oscillation continues to this day, as documented by the evolutionary psychologist Steven Pinker in *The Blank Slate*, there is increasing appreciation, perhaps especially in the arena of criminology, that both nature and nurture make important contributions. This change comes, in part, from a far greater understanding of human genetics, combined with long term studies of how humans and other animals develop within environments that are either nurturing or damaging. Understanding this variation is essential as it shapes the outcome of evil's recipe — of whether the combination of desire and denial results in gratuitous cruelty[58].

---

[58] Pulling apart nature and nurture: Alia-Klein, N., Goldstein, R., Kriplani, A., & Logan, J. (2008). Brain Monoamine Oxidase A activity predicts trait aggression. *Journal of Neuroscience, 28*(19), 5099-5104; Beaver, K.M, DeLisi, M., Vaughn, M.G., & Barnes, J. C. (2010). Monoamine oxidase A genotype is associated with gang membership and weapon use. *Comparative Psychiatry, 51*(2), 130-134; Bukholtz, J.W, & Meyer-Lindenberg, A. (2008). MAOA and the neurogenetic architecture of human aggression. *Trends in Neurosciences, 31*(3), 120-129; Derringer, J., Krueger, R.F., Irons, D.E., & Iacono, W.G. (2010). Harsh discipline, childhood sexual assault, and MAOA genotype: an investigation of main and interactive effects on diverse clinical externalizing outcomes. *Behavioral Genetics, 40*(5), 639-648; Enoch, M-A., Steer, C.D, Newman, T.K., Gibson, N., & Goldman, D. (2010). Early life stress, MAOA, and gene-environment interactions predict behavioral disinhibition in children. *Genes Brain & Behavior, 9*(1), 65-74; Fergusson, D.

Consider the MAOA gene that I mentioned in the last section. This gene produces an enzyme that goes by the same shorthand of MAOA, or **M**ono**A**mine **O**xidase **A**. MAOA is evolutionarily ancient, present in other distantly related animals, and has two different forms — *low* and *high* — that influence the level of serotonin as well as the brain areas involved in social evaluation and emotional regulation. Early evidence for the critical role

M., Boden, J. M., Horwood, L. J., Miller, A. L., & Kennedy, M. A. (2011). MAOA, abuse exposure and antisocial behaviour: 30-year longitudinal study *The British Journal of Psychiatry, 198*(6), 457–463; Gibbons A. (2004) American Association of Physical Anthropologists meeting. Tracking the evolutionary history of a "warrior" gene. *Science* 304:818; Lea, R., & Chambers, G. (2007). Monoamine oxidase, addiction, and the "warrior" gene hypothesis. *Journal of New Zealand Medical Association, 120*(1250), 1-5; Meyer-Lindenberg, A., Buckholtz, J.W., Kolachana, B., Hariri, A., Pezawas, L., Blasi, G., Weinberger, D.R. (2006). Neural mechanisms of genetic risk for impulsivity and violence in humans. *Proceedings of the National Academy of Sciences (USA) 103*(16), 6269-6274; Sebastian, C.L., Roiser, J.P., Tan, G.C.Y., Viding, E., Wood, N.W., & Blakemore, S-J. (2010). Effects of age and MAOA genotype on the neural processing of social rejection. *Genes Brain & Behavior, 9*(6), 628-637; Williams, L.M, Gatt, J.M, Kuan, S.A., Dobson-Stone, C. Palmer, D.M., Paul, R.H., Gordon, E. (2009). A polymorphism of the MAOA gene is associated with emotional brain markers and personality traits on an antisocial index. *Neuropsychopharmacology, 34*(7), 1797-1809.

of this gene in social behavior emerged from a study that knocked it out of operation in mice. The result: hyper-hyper-aggressive mice. These genetically trans-formed mice had no capacity to regulate their social behavior. Consequently, all interactions were treated as confrontational and handled by aggressive attacks. These results are consistent with a large body of work in animals showing that low levels of serotonin and heightened aggression go hand in hand. These results are also consistent with work on humans.

In 1993, the biologist Hans Brunner analyzed the genetics of a large, extended family. Some individuals within this family were born with a defect that silenced the operation of the MAOA gene, just like the experimental mice. Relative to others in the family, these individuals had a pronounced history of violence, including murder, rape, and arson. Oddly, although this work provided one of the cleanest links between genes and violence in humans, it slid under the radar of scientific attention, only to be resuscitated and enriched about ten years later.

The behavioral geneticists Avshalom Caspi and Terry Moffitt studied a large population of young boys over several years. Though boys and girls have the MAOA gene, its effect on behavior is easier to study in boys because they have only one copy whereas girls have two, one for each of their two X chromosomes. For each boy, Caspi and Moffitt collected information on whether they were raised by parents who were caring, mildly abusive, or severely abusive, and the presence and frequency of their antisocial behavior. For each boy, they also noted whether they had the low or high expressing form of MAOA.

Caspi and Moffitt's results provide a textbook example of nature's interaction with nurture. If the parents were mildly abusive, the boys with the low activity form were nine times more likely to fight, steal, bully, and defiantly break rules. For those boys with severely abusive parents and the low activity form of MAOA, 85% developed into violent, delinquent criminals. However, if the parents were caring, the different genetic forms made no difference in their child's personality or behavior.

What these findings tell us is that in humans, it makes little difference which form of MAOA you have if you grow up with nurturing parents. But if you grow up with abusive parents, your genes make all the difference in the world. Those with the low expressing form of MAOA are more likely than not to develop into delinquents, whereas those with the high expressing form are more likely than not to develop immunity. By a double dose of bad luck, one shot from the genes and one from the environment, some have a high probability of harming others.

The neuroscientist Andreas Meyer-Lindenburg provided a link between the particular form of MAOA and differences in the brain. Those with the low expressing form of MAOA, associated with relatively poor social regulation in Caspi and Moffit's results, had significantly smaller brains, specifically in regions associated with the control of emotion and social behavior — the amygdala, anterior cingulate, and (only in men) also another part of the prefrontal cortex. These individuals also had less connectivity between these regions, harking back to the importance of our connected brains for normal functioning. Less connectivity translates to less control by the frontal areas of the brain over emotion-relevant areas such as the amygdala. When individuals with the low expressing form viewed angry or fearful facial expressions, the emotionally-relevant brain areas went into hyper-drive, whereas those areas involved in regulating emotions hibernated. Thus, in contrast with individuals who have the high expressing form of MAOA, those with the low expressing form are overwhelmed by emotionally charged experiences, lacking the mental brakes to stay cool. By luck of the draw, the low expressing form of MAOA helps sculpt a child that is more likely to get angry and violent in the face of frustration and other emotional challenges, whereas the high expressing form builds a child that is walled off, immune to the same challenges.

MAOA is not only crucial in long-term human development, but also in everyday, ephemeral social interactions. In a laboratory study, an experimenter offered subjects the opportunity to earn up to $10 on a vocabulary quiz. Once they finished the quiz, they learned that an anonymous person in another room either took some of their earnings or left it alone. With this information in hand, the quiz-taker could either vindictively punish the person by giving them some hot sauce or they could cash out of the game and recover the money lost. In reality, there was no partner in the other room. When subjects with the low expressing form of MAOA lost most of their earnings, they were far more likely to deliver the hot sauce than those with the high expressing form; they were also most likely to deliver the highest amount of the sauce. Like long-term parental abuse, even short-term provocation invoked in a laboratory environment can cause those with the low expressing form of MAOA to act out and attack.

These results link back to my proposed recipe for evil by pointing out that individuals differ in their capacity to resist the temptation of desire, whether it is the desire to attain resources by violent or non-violent means. For those with a low-expressing form of the MAOA gene and a bad upbringing, their capacity for self-control is weak. As a result, they are more likely to follow up on a desire to hurt those who get in the way. This is the kind of evidence that reveals why some individuals are more at risk than others when it comes to harming others. Though the process starts with genetic differences, it is a process that is shaped by what individuals experience.

As with all genes that have different forms, the number of individuals with the low expressing form of MAOA varies by population, including different ethnic and culturally identified groups. Caucasian and Hispanic males show some of the lowest frequencies at 34 and 29 % respectively, whereas Maori, Pacific Islander, and Chinese males show the highest at 56, 61, and 77% respectively. In a study of over 1000 men, individuals with the low expressing form of MAOA were more likely to be in violent gangs, and once in gangs, were more likely to use guns and knives than individuals with the high expressing form. Variation in the frequency of these two forms is interesting as it provides the signature handiwork of natural selection. When the frequency of one form goes up, the most likely explanation is that this form benefits the individual carriers. When the frequency goes down, there is a hidden cost. In light of this teeter-tottering of frequencies, the Maori are of interest. As celebrated by many New Zealanders today, the Maori were a highly adventurous and warring people. Individuals who took risks and fiercely defended their resources were heroes. Heroes may have been carriers of the low expressing form of MAOA. Heroes often leave more offspring, who were also carriers of the low expressing form. In the Maori environment, selection may well have favored this form of the gene. The important point is that different environments will favor different frequencies of the two forms of MAOA. This helps explain both some of the causes of individual differences and the challenges we face in confronting cultures of violence that are fueled by nature and nurture.

Many other labs have followed up on Caspi and Moffitt's long term, developmental study. Most find the same relationship between the MAOA gene and anti-social behavior. Others add to this account by showing how different genes and early appearing physiological differences contribute to a highly aggressive and antisocial starting state. For example, the psychologist Alexander Strobel put subjects in a brain scanner and invited them to play a bargaining game where they could punish someone who acted unfairly to them — personal revenge — or punish someone who acted unfairly to someone else — impersonal punishment. For each subject, Strobel also collected information on a gene called COMT Met. This gene has three different forms, linked to differences in activity level in the frontal lobe of the brain, which are linked to differences in the levels of dopamine, which are linked to differences in the experience of reward. Given the different forms of COMT Met, at least part of what we experience as the feeling of reward or gratification was determined by our parents, and our parent's parents, and their parent's parents, all the way back to our ape-like cousins who evolved this gene. Strobel's question was: would genetic differences in the anticipation of sweet revenge or punishment influence the decision to punish?

When Strobel looked at the brain scans of his subjects, he found that the same circuitry was engaged for personal revenge and impersonal punishment, with significant activity in two areas that I have mentioned several times before: the striatum — a reward area — as well as the insula — an area involved in the feeling of disgust. When we detect an injustice, we feel disgusted, a feeling that may motivate our desire for retribution. The striatum finishes off the process, rewarding us for our punitive response, and wiping out the negative feeling of disgust. Importantly, individuals with the high expressing form of COMT Met, and thus, higher levels of dopamine, showed stronger activation in the striatum and were more likely to punish those who acted unfairly.

Strobel suggests that those with the high expressing form of the COMT met gene punished more because they anticipated a higher level of reward. If this explanation is right, it has profound consequences for how we think about individual participation in the policing of norms and the honey hits associated with aggression. Some people will shy away from punishment, not because they fail to see the importance of ratting out cheaters, but because they don't anticipate feeling good about it. Others are more likely to punish even the most minor infractions because they feel empowered and good about it. Those who are empowered to punish because it feels good have forged a stronger association between aggression and reward. Unbeknownst to these individuals, they started life with a bias, one that colored their willingness to harm others. This bias is joined by many others that I discuss in the next few sections.

The take home message is that if you are born male, endowed with certain genetic variants such as the low activity form of MAOA, and experience physical and psychological abuse by your parents, the odds of delinquency are frighteningly high. That's the bad news. The good news is that if you are born male, have the high activity variant of the MAOA gene, and experience physical and psychological abuse by your parents, you are vaccinated by nature against the harms of your unfortunate nurture. The problem, of course, is that you have no say over which endowment you get, nor over the kinds of parents you live with.

Marc D. Hauser

One of the reasons I have worked through this case study of genetic constraints and environmental sculpting is to provide an antidote to the often polarized views that have dominated much of the historical and psychological literatures on evil[59]. Many of the earliest, and most famous psychological experiments were related in one way or another to the philosopher Hannah Arendt's thoughts about the Nazi commander Adolf Eichmann and the fact that good people are capable of horrific things: the banality of evil. Based on Arendt's description, Eichmann was not a maniacal monster who desired the annihilation of all Jews and relished the idea of their extermination. Rather, he was merely a cog in the Nazi

---

[59] Banality, situationism and authority: Zimbardo, P.G. (2004). A situationist perspective on the psychology of evil: Understanding how good people are transformed into perpetrators. *In: The Social Psychology of Good and Evil,* A. Miller (ed), NY: Guilford Press, pp. 1-23; Haslam, S.A., & Reicher, S. (2007). Beyond the banality of evil: three dynamics of an interactionist social psychology of tyranny. *Personality Social Psychological Bulletin, 33*(5), 615-622; Haslam, S.A. & Reicher, S. (2012). Contesting the "nature" of conformity: what Milgram and Zimbardo's studies really show. *PLOS Biology* 10(11): 1-4; Russell, N. (2010). Milgram's obedience to authority experiments: Origins and early evolution. *British Journal of Social Psychology* 50: 140-162; Sabini, J., & Silver, M. (2007). Dispositional vs. situational interpretations of Milgram's obedience experiments: The fundamental attributional error. *Journal for the Theory of Social Behaviour, 13*(2), 147-154.

machine, following orders. On this view, behind every average Joe is a person equipped with an engine of malice. Banality is the veil of evil. Following this line of thinking, the psychologist Stanley Milgram famously showed that normal people were capable of shocking innocent others when an authority figure told them to do so; of course, there were no shocks, but the subjects believed they were real. Similarly, the psychologists Solomon Asch and Philip Zimbardo showed that normal people followed group attitudes and instructions, no reflection, no critical thinking, no concern about the consequences of their actions. In Zimbardo's study — the well known Stanford prison experiments — run of the mill undergraduates playing the role of prison guards turned into little dictators without any instruction, mentally and physically abusing other run of the mill undergraduates playing the role of prisoners. Together, these studies seemed to support a blank slate view of the mind, a tablet waiting for inscription by the local culture, with no constraints on the written matter. Add authority to conformity and you have a society that will relish gratuitous cruelty.

A closer look at Eichmann, as well as details of the studies by Milgram and Zimbardo, reveals a different story. Arendt ignored important details of Eichmann's life and comments. Eichmann actively sought ways to carry out the Final Solution of exterminating the Jews, often by ignoring orders from his superiors. Prior to his trial, he expressed only one regret: that he had not destroyed more Jews. This is hardly the voice of banality. In parallel, and as the psychologists Haslam and Reicher present in several important papers, there was far greater variation in how subjects responded to authority and conformity. Many subjects in both the Milgram and Zimbardo studies refused to follow the orders or rules of the game. Those who refused tended to identify more with the victim and less with the authority figure or ideology. This suggests important individual differences in the capacity to experience empathy and compassion for another. These differences directly impact whether an individual will harm another and what, if anything, they feel while doing it.

Studies by the cognitive neuroscientist Esse Viding provide an insightful entry into the source of these individual differences, showing that by the pre-school years, some children have a diminished capacity for empathy, expressing a deeply callous and unemotional character. These children exhibit severe conduct problems, especially violence. These children lack self-control, remorse and an awareness of others' distress. They are cold, heartless kids — highly similar to the adult form of psychopathy. If they have a twin, they are more likely to share this callous-unemotional personality than two unrelated children, revealing the trademark of a powerful genetic engine. More boys than girls fall on the high end of this callous-unemotional scale — where high translates to colder and more callous. Those who score highest on the scale engage in more direct physical bullying than those who score lower. High scorers lack the skills to modulate their behavior following direct or anticipated punishment. These individuals also show reduced activity in the amygdala, a part of the brain that is critically involved in regulating emotion, especially the assignment of a positive or negative value on our actions and experiences. These individual differences persist into adulthood. These are the kind of individual differences that can explain why some followed Milgram and Zimbardo's instructions to perfection, while others resisted, exerting self-control.

The sweetness of control

When humans and other animals travel the road to excess, whether for food consumption, power, sex, or violence, it is often because a history of losing self-control turned into an addictive habit of giving into temptation. What causes us to lose our sense of moderation, allow our mental brakes to slip, and give in to temptation? What causes our preferences to inconsistently and irrationally shift over time, allowing seductive offerings to win? If you are the social psychologist Roy Baumeister who has contributed fundamental insights into the nature of evil, the answer is simple: Sugar. Want it. Love it. Need it.[60]

---

[60] The science of self-control: Baumeister, R.F. (2002). Yielding to temptation: Self-control failure, impulsive purchasing, and consumer behavior. *Journal of Consumer Research, 28*(4), 670-676; Baumeister, R.F., & Alquist, J. (2009). Is there a downside to good self-control? *Self & Identity, 8*(2), 115-130. Baumeister, R.F. (2002). Yielding to temptation: self-control failure, impulsive purchasing, and consumer behavior. *Journal*

When we work hard, focusing on a difficult problem or trying to figure out the best decision, exhaustion strikes. Part of our exhaustion seems to come from depleting a critical resource: sugar, or more precisely, glucose. When the availability of this resource diminishes, we also lose self-control. This is why, as discussed by Baumeister and Tierney in their book *Willpower*, the loss of self-control has a cycle that follows the time of day,

---

of Consumer Research, 28, 670-676; DeLisi, M., & Vaughn, M. G. (2007). The Gottfredson-Hirschi critiques revisited: reconciling self-control theory, criminal careers, and career criminals. *International Journal of Offender Therapy and Comparative Criminology, 52*(5), 520–537; DeWall, C., Baumeister, R., & Stillman, T. (2007). Violence restrained: Effects of self-regulation and its depletion on aggression. *Journal of Experimental Social Psychology* 43: 62-76; Gailliot, M. T, & Baumeister, R. F. (2007). The physiology of willpower: Linking blood glucose to self-control. *Personality and Social Psychology Review* 11(4), 303-327; Muraven, M, Tice, D.M, & Baumeister, R.F. (1998). Self-control as limited resource: Regulatory depletion patterns. *Journal of Personality and Social Psychology, 74*(3), 774-789; Vohs, K. D., Baumeister, R.F., Schmeichel, B.J., Twenge, J.M., Nelson, N.M., & Tice, D.M.. (2008). Making choices impairs subsequent self-control: a limited-resource account of decision making, self-regulation, and active initiative. *Journal of Personality and Social Psychology, 94*(5), 883-898; Dewall, C.N., Deckman, T., Gailliot, M.T., & Bushman, B.J. (2011). Sweetened blood cools hot tempers: physiological self-control and aggression. *Aggressive Behavior*, 37, 73-80.

with the greatest losses occurring late rather than early: diet breakers are more likely to pig out in the evening than early in the morning; shoppers are more likely to buy impulsively as the day moves on; impulsive crimes and relapses of addiction are evening affairs; judges are more likely to dole out punishment at the end of a day in court than when they start a new day. Dozens of experiments show that if you have to exert self-control in one context it taxes your capacity to exert self-control in another. For example, if you ask subjects to avoid laughing while watching a comedy routine, avoid thinking about a white bear, or avoid eating chocolates now to have radishes later, these same subjects will squeeze a hand grip for a shorter period of time than subjects who never contended with the various self-control tasks. When you deplete your personal resources, you lose your grip, opening yourself up to binge eating, sexual promiscuity, drug relapses, and unnecessary violence.

How do we know that glucose plays this kind of role? If you give people a milkshake with real sugar before they have to take a hard test involving self-control, they do better than if you give them a milkshake with an artificial sweetener. If you first make people take a test that taxes their self-control and causes their glucose to drop, they do worse on a subsequent test, including the hand grip squeezing test. In an extraordinary series of experiments and observations, the psychologist Nathan DeWall found that subjects who drank lemonade with glucose were less likely to respond aggressively to an insult than subjects who drank lemonade with artificial sweetener; individuals with diabetes – who have difficulty regulating blood glucose, and thus have less of it — reported higher levels of aggression on a questionnaire than non-diabetics; within the United States, those states with higher numbers of diabetics showed higher crime rates; and countries with a higher frequency of a genetic disorder that lowers glucose levels showed higher killing rates both in and out of war.

To accept DeWall's striking results, it is necessary to accept one connection between glucose and self-control and a second between self-control and aggression:

GLUCOSE DOWN → SELF-CONTROL DOWN
SELF-CONTROL DOWN → AGGRESSION UP

That aggression often follows from a loss of control is backed up by considerable evidence, including clinical studies that link lack of inhibition in psychopaths to extreme violence. Also of interest is the fact that impulsive aggression is more likely to arise when individuals are drunk than sober. Alcohol, as we all know, lowers our inhibitions, but also lowers glucose in both the brain and body. Though scientists such as Baumeister and DeWall have not yet worked out in detail how glucose is used or replenished in the context of self-control[61], there are far too many studies using different methods

---

[61] see especially Hagger, M. S., & Chatzisarantis, N. L. D. (2013). The sweet taste of success: The presence of glucose in the

and subjects to ignore this relationship. Minimally, these studies suggest that we should think about self-control like a resource, something that can be used up and replenished. When it is depleted, actions that are normally suppressed are released, resulting in addictions, over-eating, and violence.

One of the interesting implications of DeWall's work for understanding the variation in our species' potential for harm, is that individual differences in glucose availability are coupled with individual differences in self-control which are, as noted, linked to differences in violence. Diabetes shows a high level of heritability, meaning that some individuals are more likely to develop this problem than others simply based on what genes they received from their parents. The prevalence of diabetes is on the rise in many countries, with some estimates suggesting that by 2025, there will be 325,000,000 diabetics world wide, more than double current estimates. The genetic disorder that lowers glucose levels arises because of a deficiency in a key enzyme, glucose-6-phosphate-dehydrogenase. This is one of the most common enzyme deficiencies in the world, affecting over 400,000,000 people, and in many cases, triggered by the consumption of fava beans. As with variation in the frequency of MAOA, so too can variation in this glucose-related gene be subject to selection pressures, especially given its link to violence. Once again, we see nature and nurture

oral cavity moderates the depletion of self-control resources. *Personality and Social Psychology Bulletin* 39(1): 28-42.

contributing to individual variation and cultural differences in our capacity to harm others.

Together, these observations of glucose-related disorders speak to a disconcerting reality: we are born with inherent differences in the availability of key resources guiding self-restraint. Some of us start off life better equipped to control our frustrations, wait for future gains, and moderate our temper. These early differences can have long lasting and disastrous effects later in life, a point supported by a famous study that began forty years ago with children sitting in front of a marshmallow.

The social psychologist Walter Mischel recruited four year old boys and girls to his laboratory and sat them down at a table with only two objects, a marshmallow and a bell. He then told each child that he was going to leave the room. If they wanted to eat the marshmallow, they only had to ring the bell. But, as Mischel informed them, if they waited for his return, he would bring them more marshmallows. Mischel took out his stopwatch and recorded how long each child waited before ringing the bell.

Some children rang the bell almost immediately, leaving Mischel no time to leave the room. Others waited. This isn't surprising. Some children are impulsive, others are patient, and these personality differences are apparent early in life. What is surprising is that these early appearing personality differences held steadfast, impacting later life decisions and actions. The more impatient types were more likely to be involved in juvenile delinquency, have poor grades, abuse drugs, get divorced, and lose their jobs.

For women who developed eating disorders, those who were more patient as children were more likely to develop anorexia, whereas those who were more impulsive were more likely to develop bulimia. When the developmental psychologist B.J. Casey put these now 40-somethings inside a brain scanner, the patient ones showed stronger activation in the prefrontal areas of the brain when viewing happy and fearful faces, revealing stronger self-control over their feelings. In contrast, when the impatient ones viewed the same stimuli, not only was there a weaker response in the prefrontal region but a stronger response in the ventral striatum when viewing happy faces. The striatum, as noted earlier, is involved in the experience of reward. For the impatient types, seeing something positive is like eating candy, something that is hard to ignore. The patient types regulate this feeling, transforming the heat of the moment into a cooler experience. The impatient types are overwhelmed by this feeling, giving into temptation. This work adds to the genetic evidence reviewed earlier, showcasing both the importance of individual differences in self-control, the stability of these differences, and their predictive power in explaining our health, wealth, and proclivities for violence.

Mischel's work suggests yet another way in which we start out life with different potentials to cause excessive harm. From giving in quickly to a marshmallow, we may also give in quickly to causing excessive pain and suffering.

Individual differences in glucose metabolism, together with relative differences in brain activity, lead to stable

differences in self-control. But there's more, both luck of
the draw genetic effects and clinical distortions. Recall
that the low expressing form of the MAOA gene results
in lower levels of serotonin which, in turn, leverages less
control over aggressive impulses. There is another gene
— SLC6A4 — that also comes in two forms and regulates
the level of serotonin. The short form of this gene gives
you less serotonin, is commonly found in pathological
gamblers and psychopaths — two heavily male-biased
disorders that are associated with impoverished impulse
control. Psychopaths also have relatively smaller frontal
lobes, especially within a region that has a high density
of serotonin neurons. Psychopathy is joined by a family of
impulse control disorders that also implicate dysfunction
of the serotonin system, including kleptomania (stealing),
pyromania (burning), trichotillomania (hair pulling), and
oniomania (shopping). Like glucose, serotonin plays a
lead role in our capacity for self-control. When serotonin is sidelined from the performance, any number of
impulsivity problems may emerge.

What I have said thus far is only a partial accounting
of the biological ingredients that figure into our capacity for
self-control. What we learn is that regardless of the situation, some individuals are better inoculated against the
pull of authority and group ideology and others are more
susceptible. If you missed the inoculation clinic in utero,
you are more susceptible to temptations and excesses,
including excessive violence. This is important for our
interpretation of the real world and of the famous psychological experiments by Milgram, Zimbardo, and others in

which seemingly good people carried out unambiguously horrid things. Some individuals carry a genetic skeleton that resists the push and pull of charismatic leaders and powerful isms. These people will not be pushed into doing bad things. Others, faced with the exact same situation, will find their skeleton buckling, tempted to take risks and lash out when the going gets tough.

Invisible risks

In real life, there are risks associated with every decision, some clear from the start and others only clear in hindsight. As with self-control, a growing body of evidence suggests that there are individual differences in risk-taking: some are risk-averse, some risk-prone, and some seemingly risk-blind, unaware that they are taking risks at all[62]. Some of these differences are evident early

---

[62] The science of risk: Fairchild G., Passamonti L., Hurford G., Hagan C.C., von dem Hagen E.A.H., van Goozen S.H.M., Goodyer I.M., Calder A.J (2011). Brain structure abnormalities in early-onset and adolescent-onset conduct disorder. *American Journal of Psychiatry* 168:624–633; Gao, Y., Baker, L. A., Raine, A., Wu, H., & Bezdjian, S. (2009). Brief Report: Interaction between social class and risky decision-making in children with psychopathic tendencies. *Journal of Adolescence, 32*(2), 409–414; Gao, Y., Raine, Adrian, Venables, P. H., Dawson, M. E., & Mednick, S. A. (2010). Association of poor childhood fear conditioning and adult crime. *American Journal of Psychiatry, 167*(1), 56–60; Glenn, A. L., Raine, A., Venables,

Marc D. Hauser

in life. Some of these differences are strongly associated with crime later in life. Some of these differences provide insights into the invisible risks that individuals confront, risks that can cause great harms.

Research on clinical populations with antisocial disorders, most notably those with a clinical diagnosis of psychopathy, reveals a major cause of their high risk, costly, and violent behavior: a failure to experience fear, anxiety, or stress in response to highly evocative images and sounds. In contrast with healthy populations, psychopaths are emotionally blasé about the things in the world that can cause harm or result in punishment. The problem lies in the fact that psychopaths, both adults and those identified as candidates early in childhood (recall Esse Viding's work mentioned in the last section), fail to learn about the dangers in life. Their failure to learn is caused by a reduction in size and activity

P. H., & Mednick, S. A. (2007). Early temperamental and psychophysiological precursors of adult psychopathic personality. *Journal of Abnormal Psychology, 116*(3), 508–518; Isen, J., Raine, A., Baker, L., Dawson, M., Bezdjian, S., & Lozano, D. I. (2010). Sex-specific association between psychopathic traits and electrodermal reactivity in children. *Journal of Abnormal Psychology, 119*(1), 216–225; Raine, A, Venables, P., Dalais, C., Mellingen, K., Reynolds, C., & Mednick, S. (2001). Early educational and health enrichment at age 3–5 years is associated with increased autonomic and central nervous system arousal and orienting at age 11 years: Evidence from the Mauritius Child Health Project. *Psychophysiology, 38*, 254–266.

of two critical and connected brain areas: a region of the frontal cortex and the amygdala. When this system works efficiently, it allows individuals to learn about the sounds, smells, and sights that are associated with bad things in the world. When this system works well, individuals learn to avoid antisocial, immoral, and illegal acts by developing anxiety and fear over the possibility of punishment and personal injury. When this system works poorly, as is the case in psychopaths, individuals act as if there are no dangers or risks of punishment — a disposition that enables inappropriate actions. But, as noted earlier, psychopathy covers a broad spectrum, with problems that all of us confront at some point in our lives, some of us even repeatedly. This is important as it forces us to look at non-clinical populations for the causes of individual differences in risk-taking, especially our reactivity to dangerous events.

Studies carried out over several decades, led by the developmental psychologist Jerome Kagan, reveal that children in a variety of cultures begin life with distinctive temperaments. Some are mellow, blasé about events that are startling to many. Others are high strung and reactive, responding with heightened fear to the same startling events. Others fall somewhere in between these two poles. What is surprising is the fact that those with the flattest response to evocative images and sounds are the most likely to become violent delinquents in young adulthood.

In a remarkable study, the neuroscientist Adrian Raine and his colleagues presented 1,795 three year-olds with two different sounds while recording the sweatiness of their palms; the sweatier the palms, the greater the stress and fear. One sound was always associated with a second and highly aversive noise, while the second sound was always played alone. When you pair a neutral sound, such as a pure tone, with a nasty sound such as fingernails on a blackboard, simply hearing the pure tone will make your skin crawl; the pure tone predicts what is coming, and what is coming is not pleasant. When Raine revisited these same individuals twenty years later, those with serious criminal records (drug abuse, dangerous driving violations, or violence) had the driest palms at the age of three years. No stress. In another study, focusing specifically on violence, Raine measured the same stress response in a different group of three year-olds and then looked at their level of violence five years later. Once again, those with the driest palms at three years were the most violent at eight years. In the absence of a system that enables individuals to learn about danger, the brain and body act as if they were shrouded in an invisibility cloak, blind to the risks of crossing either moral or legal lines.

Raine's findings fit well with the marshmallow study. In the same way that those who were most impatient in the pre-school years were also most likely to exhibit signs of delinquency in early adulthood, so too were those who were most blasé about fearful stimuli as children most likely to exhibit delinquency in adulthood. Both studies reveal the stability of personality traits created early in development. Both studies suggest that at the level of groups of individuals, as opposed to specific individuals, the blasé-impatient types represent a greater threat to our welfare. The point here about groups is important. These studies do not allow us to look at an individual's record and conclude that because he could only wait for 3 seconds before eating the lone marshmallow, and almost fell asleep when presented with loud banging noises, that he is without doubt headed for a life of crime. We also can't conclude that because patience and reactivity to fearful stimuli can be measured as early as three years old, that these personality traits are entirely genetic and fixed. In fact, other studies carried out by Raine show that if you ramp up the nutrition, exercise, and mental stimulation of children between the ages of 3-5 years, you can reduce adult criminal offenses by 35%. This reveals the importance of experience and the plasticity of this system over development.

What we can conclude from these findings is that there are significant individual differences that affect who is willing to take risks and who isn't. We can conclude that there is a strong biological component that constrains the individual's options. We can conclude that those who start early in life without an understanding of the dangers in the world, act as if they live in a risk-free world. Molecular biologists provide an increasingly precise understanding of how these individual differences start, pointing to genes that bias some individuals to take extreme risks, including the risk of violating social norms and laws by violently attacking another human being.

There are many situations where taking a risk pays off, whether we think of stealth military operations, chancy shots in the final seconds of a basketball game, or significant investments in an up and coming stock option. Though it often pays off to play it safe, those who stick their necks out and take a chance, may bring home significant gains. It is because of these competing strategies and potential payoffs that evolutionary biologists have imagined that selection could maintain both personality types within a population — a point noted earlier for the MAOA and glucose-related genes. If selection has worked in this way, then there must be genetic variation that allows for both strategies. To date, the strongest evidence comes from a family of genes associated with the regulation of dopamine, with the memorable acronyms of DAT1, DRD2 and DRD4; each of these genes is associated with different forms, each form associated with the expression of different levels of dopamine. Recall from earlier chapters that dopamine plays an essential role in our experience of reward, including how motivated we are to get it and what we anticipate based on our understanding of the situation — have we been rewarded in the past, how often, and how much? The idea here is that those who carry genes that output a higher level of dopamine may weight rewards more heavily and thus, show risk-blindness; for these individuals, the eye is on the prize, not the path or obstacles to this prize.

Across a number of studies, results show that variation in the expression of these genes is associated with
high-risk and low self-control, including pathological
gambling, substance abuse, sensation seeking, and
financial investments. For example, in two separate
studies, individuals with different variants of the DRD4
gene played a financial investment game involving real
money. In one, designed by the psychologist Joan Chiao,
subjects decided to invest in either a risky asset with
variable returns or a riskless asset with consistent returns.
In the second study, the economist Ana Dreber and
the evolutionary psychologist Corin Apicella allowed
subjects to either walk away with an initial starting pot
of money, or to invest some of it in a risky asset. Both
studies revealed that those with the DRD4 variant that
expresses higher levels of dopamine were more likely
to pursue the risky investment.

This work suggests that part of the variation we
observe among people who make risky investments,
drink too much alcohol, or gamble with their income, is
influenced by variation in the dopamine family of genes.
These are hidden risks that come to life thanks to the
molecular biologist's microscope. What also comes
to life is the fact that these same genes are relevant to
violence, tilting some further in the direction of striking
out even though there are significant risks and horrific
consequences.

In several studies, using an American health data
base of several thousand adolescents, results consistently
show a relationship between particular variants of the

dopamine genes and violence. For example, the sociologist Guang Guo examined the relationship between violent delinquency — involving use of guns and knives — and variation in DRD2 and DAT1 among 2,5000 individuals ages 12-23 years. DRD2 was of particular interest because medical records and clinical trials reveal that administering haloperidol — a DRD2 dopamine antagonist — helps control aggression in psychotic patients. Guo found that levels of violence were about twice as high for one variant of the DRD2 gene than others, and about 20% higher for a particular variant of the DAT1 gene. These genetic variants lead to relatively higher levels of dopamine, which leads to differences in expected and experienced reward, which leads to differences in perceived risk, which leads to differences in the odds of getting in a fight and harming others. These are not genes for aggression, violence or evil. There are no such genes. Rather, they are genes that change our perception of risk. Because risk is related to all sorts of decisions, these genes can affect the odds that we directly harm others. They are part of the story of individual differences, and part of the story of why some are more likely to engage in harming others.

I don't feel your pain

Soon after I entered junior high school, some time around my fourteenth birthday, I had a few near-death experiences, coming very close to drowning in our community pool. This was not because I was a poor swimmer. A boy named Lionel James, who was the same age but twice my size, was shoving my head under water, roaring with laughter as I struggled to gasp some air.

I usually managed to avoid Lionel in the pool, but sometimes he got the best of me while I was playing with friends. Lionel wasn't the only one who bullied me in junior high school. He was part of an evil three pack, including Ronnie Paxton and Chris Joffe, each much stronger than I. Almost daily they bruised my arms by giving me knuckle-punches, black-and-blued my chest by twisting my nipples, and locked me inside one of the school's lockers. This was no fun for me. For James, Joffe and Paxton it was delicious enjoyment.

One day my mother noticed the bruises. Horrified, she asked what happened. I reluctantly told her the story. She said we were going to see the school's principal. I told her I would prefer water drip torture. She understood and we never went to see him.

The person who rescued me from my misery was my father, a man who lived through the war as a child, running from village to village to escape the Nazis, and in so doing, confronted thuggish farm boys whose weight far exceeded their IQ. My father, upon hearing that I didn't want to go to school, offered a compromise: he would pick me up for lunch every day if I kept going to classes. I agreed, relishing the idea of escaping the lunch-time scene at school where James, Joffe, and Paxton pummeled me at will.

A month passed. I felt better. My father told me that it was time to go back to lunch at school, but with a plan, one centered around the notion of

Marc D. Hauser

respect. The only way to command it from my tormenters was to fight back. It seemed like a remarkably stupid idea. But my father was wise. I decided to give it a shot.

Days after returning to school, I found myself standing behind Paxton who displayed biceps bigger than my head. I figured I had only one shot. I tapped him on the shoulder and swung as hard as I could, hitting him square in the chest. What aim. What perfection. What wasted energy. With no more than a flinch, Paxton looked down at me, fury in his face, and grunted "What's up with you?" With tears running and lips trembling, I sputtered "I can't take it anymore. You, Joffe, and James are constantly pummeling me. I can't take it!" And then, as if his entire brain had been rewired, serotonin surging to provide self-control, dopamine flowing to shift his sense of reward, the hulk spoke: "Really? Okay, we'll stop." And just like that, James, Joffe and Paxton stopped. No more bruises, no more locker games. From victim to victory.

I was fortunate. Many are not. In the United States alone, an estimated 2 million children bully some 3 million victims per year, and of these victims, close to 200,000 will avoid school to avoid bullying, and 1 out of every 10 children will drop out of school completely. Worldwide statistics suggest that up to 20-30% of all children are bullied, and as a result suffer extreme physical and psychological trauma, including bullycide — the termination of a life that is treated by others as though it is no longer worth living. These are horrific numbers, a wake-up call that has finally caused bullying to enter the international arena of discussion on education reform and cyberspace policing.

That there are bullies, victims, victims turned bullies, bullies turned victims, and individuals who are neither bullies nor victims, showcases considerable variation in behavior and the psychology that leads to it. Common lore has it that bullies are clueless social oafs, whereas victims are geeks, short, overweight, disabled, or more generally abnormal in some way. Though some bullies are indeed social Neanderthals, while some victims are awkward or size-challenged, as I was, this crude assessment covers up a far more interesting story that has emerged from the sciences. This is a story about individual differences that, like the others I have recounted in this chapter, helps us understand why some are attracted to harming others while others are attracted to compassion, or at least non-violence[63].

---

[63] Empathy, morality and violence: Beitchman, J., Zai, C., Muir, K., Berall, L., Nowrouzi, B., Choi, E. & Kennedy, J. (2012). Childhood aggression, callous-unemotional traits and oxytocin genes. *European Child & Adolescent Psychiatry* 21: 125-132; Gini, G., Albier, P., Benelli, B., & Altoe, G. (2007). Does empathy predict adolescents' bullying and defending behavior? *Aggressive Behavior, 33*, 467–476; Gini, G., & Pozzoli, T. (2009). Association between bullying and psychosomatic problems: A Meta-analysis. *Pediatrics, 123*(3), 1059–1065; Muñoz, L. C., Qualter, P., & Padgett, G. (2011). Empathy and bullying: exploring the influence of callous-unemotional traits. *Journal of Child*

The psychologist Gianluca Gini explored the moral psychology of over 700 children between the ages of 9-13 years, classified by peers and teachers into three personality types: bullies, victims, and defenders; defenders are individuals who have never bullied or been bullied, but often step in to protect the victims. All of the children read different moral scenarios and for each, judged whether the protagonist was very bad or very good, as well as how they or someone else would feel about harming or helping another. The first set of judgments focused on whether children understand that moral decisions

---

Psychiatry and Human Development. 42:183-196; Obermann, M.-L. (2011). Moral disengagement in self-reported and peer-nominated school bullying. *Aggressive Behavior, 37*(2), 133–144; Pozzoli, T., Gini, G., & Vieno, A. (2012). Individual and class moral disengagement in bullying among elementary school children. *Aggressive Behavior,* 38(5): 378-388; Sugden, K., Arseneault, L., Harrington, H. L., Moffitt, T. E., Williams, B., & Caspi, A. (2010). Serotonin transporter gene moderates the development of emotional problems among children following bullying victimization. *Journal of the American Academy of Child & Adolescent Psychiatry,* 49: 830-840; Viding, E., Blair, R.J.R., Moffitt, T.E., & Plomin, R. (2005) Strong genetic risk for psychopathic syndrome in children. *J Child Psychology and Psychiatry* 46:592–597; Viding, E., Simmonds, E., Petrides, K., & Frederickson, N. (2009). The contribution of callous-unemotional traits and conduct problems to bullying in early adolescence. *Journal of Child Psychology and Psychiatry,* 50(4), 471–481.

depend upon what people believe and intend, as well as the outcomes that result from particular actions. The second set of judgments focused on whether children express compassion or empathy for others, and have a moral conscience with respect to their actions. For example, children might read a story about a boy who swings a bat at a baseball, but accidently hits another child in the head. Since the harm occurred accidentally, the boy is certainly not as bad as if he had swung intentionally. But regardless of the boy's intent, one should empathize with the injured child, imagining what it would be like to be him and caring about his wellbeing.

Gini's results provide a sound rejection of the social oaf explanation for bullying: bullies, along with defenders, are mature and highly competent judges of moral scenarios, piecing together information about beliefs, intentions and outcomes. Bullies recognize that a person who intentionally hits someone with a bat is morally worse than someone who does so accidentally. In contrast, victims are clueless, focused almost exclusively on outcomes: if a boy hits someone with a bat, what counts is that the person was hurt, not what the boy intended. This contrast between bullies and victims flips around when we consider the compassion and conscience side of the story: like defenders, victims express compassion and concern for those who have been hurt; bullies don't care, and nor do they feel that others should express concern.

One reason bullies don't care, or at least act as if they don't, is because they frequently disengage from the moral norms that surround them. As discussed in chapter 2, moral disengagement is a form of denial. Morally disengaged individuals are more likely to engage in harming others as they either feel justified or feel nothing at all. As several studies now show, bullies are far more likely than victims or defenders to be morally disengaged. When bullies disengage, they are most likely to justify their aggressive actions on the basis of some *higher* morality, or to downplay their actions by pointing to more egregious ones. Either way, severing themselves from their society's norms enables unconstrained violence.

HEADLINE: Bullies are morally competent but don't care, victims are morally incompetent but care, and defenders are moral saints. These results showcase important individual differences, differences that I briefly alluded to in previous sections. Esse Viding's work suggests that early in life, some children are extremely callous and unemotional, showing little compassion or empathy toward others. Some of these children maintain this personality characteristic into adulthood, exhibiting all of the traits of psychopathy, including cold, calculating, cruelty. Not only do these individuals lack empathy, but they don't care about the consequences of their own or others' actions. The important point here is that the lack of empathy is not sufficient to explain anti-social and especially violent and harmful actions. When bullies or psychopaths strike out and harm others, not only are their feelings for others flattened, so too is their concern for what others value, including life itself.

The reason I have emphasized the distinction between the capacity for empathy and the capacity to care is because of a tendency to assume that individual differences in empathy, including the virtual lack of empathy in some, is what accounts for the most significant cases of gratuitous cruelty. It's easy, so the argument goes, to torture others if you have no compassion, no feeling of what it is like to be in someone else's shoes, suffering. I have no doubt that a person with flattened empathy is capable of excessive harm. But flattened empathy isn't sufficient to explain the variety of ways in which humans engage in gratuitous cruelty, nor can it explain why some without this emotional capacity are perfectly peaceful and kind.

In Simon Baron-Cohen's recent treatment of the science of evil, empathy functionally runs the argument. Baron-Cohen makes the important point that although individuals clinically diagnosed with psychopathy and autism are often classified as such due to their lack of empathy, these two disorders result in extremely different behavioral profiles. Psychopaths have no difficulty understanding what others believe, want and intend. In fact, they are extremely good at it, which is what makes them dangerous. Knowing what others will feel and what they want, allows psychopaths to manipulate their victims, putting them in a state of fear often before ending their lives. The psychopath's problem is not so much that they lack empathy, but that they don't care about others and have an impoverished capacity to control their actions. If you don't care about others' emotions and values and lack self-control, you are not only morally disengaged but free to do as you please. Here, desire and denial have run amuck due to significant problems of brain function. In contrast, for those who suffer from autism, their lack of empathy is coupled with a lack of understanding of others' beliefs, desires and intentions, but violence does not follow. In contrast with the psychopathic condition, the autistic mind can and does express compassion for others. Autistics care in a way that psychopaths don't. Lack of empathy is therefore not sufficient to explain cruelty in either the pathological case or, I suggest, the normal case.

When we turn to cases where apparently healthy people commit gratuitous acts of cruelty, the lack of empathy not only fails to fully account for their behavior, but the argument is actually backwards: many evildoers recognize what counts as pain and suffering, have well developed empathy when it comes to their family members and friends, but explicitly develop tactics to maximize the suffering and fear experienced by others outside their inner circle. These individuals know what it is like to live in a state of fear, often because they or their relatives lived through something like this. When I described the Shawnee massacre in the last chapter, I was not describing a group of people who lacked empathy. They undoubtedly expressed it in the context of other Shawnee Indians. But when they attacked the Greathouse family it was revenge — a brutal response to having their own people massacred. When the Serbian leaders plotted their genocide, they developed an explicit plan to inject the greatest fear into Croats and Muslims, knowing what it is like to sit and watch as spouses are raped and children are brutalized. But those who carry out these atrocities are not zeroed out on the empathy scale. They have empathy, but choose to use it selectively. They have empathy but choose not to care in certain situations. Though there are numerous reasons why they don't care, they typically link to cases of revenge, the feeling — imagined or real — that they are in danger, or because selfish desires have enabled a psychology of denial that facilitates gratuitous acts of cruelty.

The boiling point

We begin life with differences in our biology, differences that place constraints on who we are and who we can become. Nothing in our biology fixes or predetermines our destiny, but as numerous studies show, differences in genetic constitution can both limit or enhance our potential for change as we grow older and interact with our culture. This variation feeds the process of natural selection and is thus directly relevant to our evolution, both its historical past and potential future. It is directly relevant to our understanding of each individual's potential for carrying out acts of gratuitous cruelty.

What is particularly striking is the observation that some of the biological differences that start each of us off in life are stubbornly resistant to modification by experience, though certainly not immune. As discussed, our capacity as young children to fend off the temptation to eat one marshmallow in favor of a future bounty predicts our salaries, physical health, drug abuse, and criminal records. Similarly, our stress response as children to an obnoxious sound sets up the odds of our committing a crime. The predictive power of these measures reveals that our biology has a remarkably strong grip on future outcomes. This perspective has important implications: what should we do with the knowledge that some children start out life with a high probability of harming others as well as themselves? What do we do with the fact that some children are closer to the boiling point than others? Not intervening seems irresponsible, while intervening seems risky given that we can't predict with certainty that a child who grabs the marshmallow immediately and is relaxed in response to an obnoxious sound will, with complete certainty, turn out to be a low income, unhealthy, drug abusing criminal. As much of this research suggests, the early warning signs are just that, signals to caretakers that children are at-risk and should be both monitored and nurtured.

Esse Viding and James Blair's research on the callous-unemotional personality dimension provides important insights into this applied issue. Those children with this personality trait are not only cold, uncaring and often aggressive, but like adult psychopaths, are also insensitive to punishment. Like adult psychopaths, young children with the callous-unemotional profile often repeat morally inappropriate actions despite a history of being punished for such actions. And like adult psychopaths, they also fail to shift their strategies in a laboratory game where different choices are associated with rewards as opposed to penalties: unlike emotionally healthy children, these children continue to make choices associated with penalties and thus, end up losing the game. One explanation for this pattern of results, shared among children and adults, is that a flattened emotional profile reduces the effectiveness of punishment. When healthy children and adults do something wrong and are punished for it, punishment hurts, bringing with it feelings of guilt and shame. For the callous-unemotional child and the adult psychopath, the emotional connection is severed. Not only is there no fear or anxiety about doing the wrong thing and being punished for it, there is functionally no awareness of the problem at all. The lack of emotions enables moral disengagement. Morally disengaged, living in a constant state of denial, their desires operate without constraint. Harming others leaves them cold. These results suggest both that early diagnosis is critical and that punishment is unlikely to work as an intervention. Though no one has yet worked out a procedure to transition children from callous and unemotional to empathic, starting early in life is the best prospect for change given the brain's heightened plasticity.

The scientific evidence on individual differences suggests that we are born with different propensities for cruelty. These propensities did not evolve for cruelty, but rather, for different aspects of social life, including the important decisions we make to survive and reproduce. Individual differences in our capacity for self-control, experience of reward, willingness to take risks, response to stress, and ability to empathize have significant biological origins. Though it is difficult to pinpoint the original function of these capacities, they play an important role today in eating, mating, playing, defending, and killing. These differences are not noise in the system, but highly relevant to our evolutionary past and futures. Individuals who were impulsive, fearless, and aggressive were invaluable when fighting against enemy tribes, and they are valuable today in modern warfare. Individuals who were patient and anticipated great rewards from building up large cattle herds, were better able to provide for themselves and their families. But these same qualities were also deployed for less virtuous goals. Many of these cases, though despicable, are not difficult to explain once we look to individual desire to acquire or maintain power. Many of these cases are, however, more puzzling as the harm caused is extreme and seemingly unnecessary to achieve the targeted end. Some are cruel for cruelty's sake. Some are cruel to intimidate the enemy. Individual differences push some of us toward gratuitous cruelty and others away from it, despite similarities in our upbringing and cultural norms.

Recommended Books

Baron-Cohen, S. (2011). *The Science of Evil*. New York: Basic Books.

Baumeister, R. & Tierney, J. (2011). *Willpower*. New York: Penguin Press.

Browning, C. R. (1992). *Ordinary Men: Reserve Police Battalion 101 and the Final Solution*. New York: Harper Collins.

Chagnon, N. (2013). *Noble Savages*. New York: Simon & Shuster.

Goldhagen, Daniel Jonah. (2009). *Worse Than War*. New York: Public Affairs.

Jones, A. (2010). *Genocide: A Comprehensive Introduction*. New York: Routledge.

Kagan, J. (2010). *The Temperamental Thread*. Boston: Dana Press.

Kiernan, B. (2007). *Blood and Soil: A World History of Genocide and Extermination from Sparta to Darfur*. New Haven: Yale University Press.

Milgram, S. (1974). *Obedience to Authority*. New York: Harper & Row Publishers, Inc.

McCullough, M. (2008). *Beyond Revenge: The Evolution of the Forgiveness Instinct*. New York: Jossey-Bass.

Singer, P. (2010). *The Life You Can Save*. New York: Random House.

Zimbardo, P. (2007). *The Lucifer Effect: Understanding How Good People Turn Evil*. New York: Random House.

epilogue: evilightenment

Educate your children to self-control, to the habit of holding passion and prejudice and evil tendencies subject to an upright and reasoning will, and you have done much to abolish misery from their future and crimes from society.

– Benjamin Franklin

Charles Darwin observed that of all the differences between humans and other animals, one capacity reigns supreme: we alone have the ability to contemplate what others *ought* to do. We alone are endowed with a moral imperative to reflect, consider, and imagine alternatives. We alone are impelled to be dissatisfied with the status quo, urged to contemplate what could be and ultimately what must be. This capacity creates a fundamental principle of human existence and enlightenment: we alone invest in the survival of the *least* fit. We give money to those in abject poverty, risk our lives to help others in areas of conflict, adopt abandoned children, nurture individuals with extreme disabilities, and care for the elderly. This principle fuels our humanitarian efforts. Sadly, it is often a necessary response to another unique difference between humans and other animals: we are the only cruel animal. We alone are responsible for creating work for those in the humanitarian sector, terrorizing innocent victims, inflicting unimaginable pain and suffering, and often, annihilating massive swaths of society. We alone are evil. The mystery is why normal, healthy individuals are gratuitously cruel, often for no apparent benefit. This is the mystery I have attempted to explain. What I have suggested is that we can explain evil by looking both to the unique evolutionary changes in our brain's design, as well as to the simple recipe that some follow as they attempt to satisfy addictive desires by means of denying reality. Here I take stock of these ideas and then draw some implications.

In the beginning, before there were bald, bipedal, big-brained, babbling humans, there were hairy, quadrupedal, bitsy-brained, barking bonobos. These animals, clearly clever, have survived for over 6-7 million years, despite attempts by our species to demolish their habitat. But — and this is a significant *but* — in the millions of years that encompass their evolutionary history, bonobos have remained virtually unchanged. They are still hairy, quad-rupedal, bitsy-brained and barking. They still live in the jungles of Africa. Not a single bonobo, or its close relative the chimpanzee, has ever taken a step out of Africa the way that members of our species did some 60-100,000 years ago. In fact, not a single bonobo or chimpanzee has ever ventured across national borders within the continent to explore new opportunities or develop new cultures. Not a single bonobo or chimpanzee has even moved out of the forests and on to the beaches or deserts or alpine environments of Africa. Not one. When we took our steps out of Africa, we did so with confidence, ready to tackle new environments, create novel tools, engage in rituals to commemorate the dead, build fires to cook food and keep warm, join hands with unrelated strangers in the service of cooperation, and create oral histories that could be passed on to generations of children. What enabled this celebratory migration was a cerebral migration. Not only did our brain get much bigger than the one housed within bonobo and chimpanzee skulls, it evolved into an engine that generates an unlimited combination of thoughts and feelings. We uniquely evolved a combinatorial brain.

What did our uniquely designed brain buy us as we started our planetary sprawl? In a word: "creativity." It enabled regions of the brain that evolved for highly specialized functions to intermingle with other regions of the brain to create new ways of thinking about and expressing what we see, hear, touch, taste and feel. A combinatorial brain paved the way for awe-inspiring bursts of creativity in art, music, literature and science, as well as in the organization of society, including its laws and governing bodies. A combinatorial brain enabled us to imagine things we have never directly experienced, to create once unimaginable worlds, including blissful heavens and living hells. My focus in this book has been the infernos we create for other human beings, here on this earth. What I have argued is that we got here as an incidental consequence of our brain's creativity. Once in play, however, this capacity was used strategically. By threatening to carry out seemingly crazy, irrational and unpredictable acts of gratuitous cruelty, we intimidate the enemy, often to a state of absolute passivity. By carrying out over-the-top, energetically wasteful acts of violence, we send an honest signal of power to our competitors. Those with the resources to waste can waste them on such gratuitous acts of cruelty. From a mere incidental consequence of connecting up different brain regions, we were handed an ability to impress with excess. These ideas speak to the evolution of evildoers, and the material covered in part II.

In part I, I suggested that evildoers are born when unsatisfied desires accumulate and combine with the denial of reality. In some cases, this combination causes only personal damage, as in all of the addictions to food, alcohol, drugs, gambling, and shopping. In other cases, this combination causes damage to others without leaving a single emotional scar on the perpetrator. In fact, most perpetrators feel good about their actions. Sometimes, this feel-good feeling is aided by moral disengagement, a state of mind that not only facilitates harming others by hibernating our moral standards, but actively encourages violence in the service of dispensing with society's parasites and vermin. Sometimes this feel-good feeling is aided by self-deception, a form of mind distortion that enables us to perceive those outside the inner sanctum as a threat to our sacred values. However we get to the space where our capacity for violence is unconstrained, we develop desires to hurt others that, like an addiction, is never satisfied. Though painlessly punishing cheaters or killing the enemy has always been an option for our species, we have often found ways of inflicting excessive pain and suffering first. This is excessive violence caused by normal human beings, not madmen. This is gratuitous cruelty that some enjoy. This is evilicious.

What can we do? How can we harness our understanding of evil to predict when it might occur again? Can we reduce future danger? Though these questions tilt us away from the descriptive level that I have pursued in this book, I can't resist dipping my toe into these prescriptive waters as we end this journey into evil. I can't resist some discussion of what we *ought* to do — our uniquely human imperative.

All legal systems impose age-related restrictions on when we can do different things. These restrictions are commonly designed to minimize future risks to both the individual and others that he or she may encounter, or to guarantee some minimum level of experience for the job. The specific age limits assigned are, however, puzzling. Why do we allow 16 year olds to drive in many parts of the United States, but prevent them from drinking alcohol until 21 and from renting a car until 25? Why must the President of the United States be at least 35 years old, but members of the House of Representatives can enter at 25? If 16 is the magic number for driving, why isn't it also the magic number for drinking, voting, becoming president, marrying without parental consent, joining the military, and being executed for a felony murder? Or why not make 21 the magic age for all age-restricted behaviors and positions? This would make sense in terms of our biology: it is precisely around the age of 21 that our frontal lobes have matured more completely, thereby providing us with a more functional engine for self-control. Or, why not question why we have a legal age at all? Why not have a brain scan for frontal lobe maturation along with a test for self-control that would allow some pre-16 year olds to drive, but might prevent some post-21 year olds from drinking? If experience is relevant, as in the case of the executive branch of the government, why not establish the nature of such experience, which might allow a brilliant and experienced 21 year old to become President in favor of a less experienced 50 year old. And if you are in favor of the death penalty — I'm not — than why not detach it from age altogether and look at the individual's moral competence and capacity for self-control?

These are hard questions. How we answer them will have resounding implications for law and society. Increasingly, legal arguments are premised on scientific evidence, as revealed by the US Supreme Court Decision in 2012 to end life without parole for youths who committed murder: much of the argument was based on the mind sciences and evidence of changes in brain maturation over development. When a legal system decides that someone can drive, drink, vote, kill, run for president, marry, and die as a penalty for crime, it has constrained human behavior based on a statistical evaluation of psychological capacity. In each case, our assignment of age-appropriateness indicates that we believe the person is responsible for his or her actions and thus, his or her future actions. It also indicates that those under age are not responsible for their actions. We grant permission to drive at 16 years of age because we believe that *most* 16 years olds are capable of driving responsibly, imposing minimal risks to others, now and in the future. We believe that a person who committed a heinous crime at the age of 18 is responsible and is likely to repeat this crime or worse in the future. He or she is thus eligible for the death penalty, at least in some states within the United States. In contrast, we believe that a 17 year old is still developing and has the potential to change. In this sense, we hold them less responsible for their actions, and hold out hope that they can change; this was the basis of the Supreme Court's decision mentioned above.

Looking out at the tapestry of age-limited restrictions reveals a rather incoherent structure. In many of these cases, the cut-off age seems both arbitrary and inappropriate given the statistics. Consider the legal driving age of 16, a decision that carries with it the implication that this is a responsible and safe age group. Accident records, however, say something different. Sixteen year olds have higher crash rates than any other age group in the United States, are more likely to die in a car crash than the average of all other age groups, and car crashes are the leading cause of death among 16 year olds. North Dakotans believe that 14 year olds can drive a car. They may have fewer drivers on the road then other states, but that doesn't mean that a 14 year old won't hit them or drive off the road after irresponsibly drinking. Why not keep all youths off the road until 21 when the statistics on fatal car crashes drop? Or why not follow the lead of car rental agencies and wait for the 25th birthday?

There are at least two common answers to the driving age problem, both utilitarian or outcome-based: in farming communities and other environments where children work with their parents, it is essential to have them driving as soon as possible; and throughout the country, many parents look forward to the day when their children can drive, thereby alleviating the need for their private chauffeur service. There is no question that these are benefits. But if the cost is death to the child and others, the economics just don't work out. One option would be to lower the legal driving age for those communities or situations in which parents demonstrate the significance of young children driving for their financial security and well being. Those without this justification must wait until they are 21, frontal lobes matured and the novelty of intoxication lowered.

Given my focus on violence and harm to innocent others, the most interesting and relevant age-related issue is when someone is treated as an adult as opposed to a juvenile criminal. Within the United States, most states set the bar at 18 years, but some as young as 16. Where a state sets its bar determines whether or not the individual is eligible for the death penalty or a life sentence, as well as a host of social services. Many states with the bar currently set below 18, are presently debating whether the age limit should be raised. For some, the issue is simply one of parity: this is not an issue where states should differ, and thus everyone should be with the majority at 18 years. Others add to this discussion by arguing that it should be 18 because of brain maturation. Although it is certainly the case that a more mature brain enables better self-control and less sensation-seeking or risk-taking, there is no evidence of a reliable difference between 16, 17 and 18 year olds. Some 16 year olds are remarkably patient and risk-averse whereas some 18 years olds are remarkably impulsive and risk-prone. If this is to be a meaningful discussion about future risks, plasticity, and the opportunity for rehabilitation, it will have to grapple with the scientific evidence that is presently on offer. When we use age to distinguish between legally permissible and forbidden actions, we have acknowledged that our biology and upbringing represent mitigating factors. We believe that juvenile crimes are forgivable and their actions correctable. In fact, their crimes are forgivable because their actions are correctable. Once we admit nature and nurture into the legal calculus concerning our youths, we must also allow such factors to guide our decisions about adults with developmental disorders, brain damage, and different genetic make-up. Yet, the law seems to have a double standard: youths lack free will, whereas adults have it, even if it is somewhat diminished. But if we believe that juveniles lack a sufficiently mature capacity for self-control, planning and thinking about alternative options, then we must recognize that fully mature adults can lose these capacities as they naturally age, and can lose them at any age if they suffer from brain damage. We must also wrestle with the fact that some people are born with a genetic constitution that biases them in the direction of addiction, sensation-seeking, violence, and a lack of compassion. Perhaps they too should be banned from driving, voting, drinking, marrying and military combat. When do we look at the excessive harms caused by individuals and hold them responsible? When do we punish them to pay for their crimes and fend off future atrocities?

The law often invokes the notion of future dangerousness as a means of evaluating risk. The general presumption is that for certain kinds of offences, there is a predictably high level of recidivism, of doing the same thing over and over again. But the implication of this judgment is that those who are deemed guilty are, in some way, not responsible for their future. Their future is determined for them. In fact, it is so determined that the law is willing to make a confident wager and send these criminals to prison or to their death. On this view, someone who has already repeated a crime is more likely to repeat than someone who has only committed a crime once. On this view, those who engage in certain kinds of crimes are more likely to repeat because it is "in" their system. Unfortunately, both folk perception and legal analysis of future dangerousness are based on weak evidence, unfounded assumptions, or both.

Consider sexual offenders. Their crime is intentional, frequently repeated, and aimed at innocent victims. Given that many sexual offenders repeat their offenses, it has the appearance of inevitability, of a process that is highly determined. Because many sexual offenders were abused as children, some experts conclude that we should blame their parents. Other experts believe that particular situations either promote or support sexual offenders, including the church and medical exam rooms. And yet other experts, including the clinical neuroscientist Boris Schiffer, reveal brain differences among pedophiles, including especially the areas involved in reward and self-control. Together, these observations suggest that the combination of a deviant nature and toxic nurture have led to a more deterministic universe.

If this assessment of sexual offenders is right, how should we think about responsibility, blame, and punishment? If sexual offenders can't help themselves, how should we assign blame? How should we assign an appropriate level and form of punishment, if punishment is even appropriate? Studies of recidivism among sexual offenders generate rates as low as 15% and as high as 80%. These studies also reveal that recidivism rates differ for incest perpetrators, rapists, and child molesters. These numbers tell us that even child molesters don't always repeat their crimes. They also tell us that sexual offences should not be lumped, but split apart into their underlying causes and triggers. Like the high odds favoring a horse with a distinguished lineage and top rated jockey, there are high odds favoring repeated sexual molestation in an individual who was sexually abused as a child and enters the clergy. Does this mean that we should all bet on this one horse or forget the race altogether? Does this mean that we should lock up the priest before he has an opportunity to enter his parish? No and No. Neither horse racing nor sexual molestation are that easily determined. Future success and future dangerousness are probabilistic. They represent our best guesses based on the available evidence. When law enforcers determine that someone is at high risk of committing a future offense, they don't care whether the individual is perfectly healthy or brain damaged. They care about risk. In terms of blame and punishment, however, the law cares about the perpetrator's brain. The law cares about a person's capacity to act rationally and independently. It is this capacity that allows us to assign responsibility. It is this capacity that drives many theories of blame and punishment, including the legal scholar Michael Moore's treatise *Placing Blame*. These are some of the reasons why scientific understanding of future dangerousness is important for law and society.

For the law to evolve, however, we need better tools to evaluate the biological underpinnings of diminished capacity. These measures, still in the early stages of development, will help refine our understanding of risk, guide our clinical interventions, and contribute to the construction of a safer society.

As we move forward, we must also recognize the rapidly changing landscape and the future dangerousness of globalization, especially its capacity to nurture individuals with a taste for harming others. Like authority, conformity, dehumanization, and self-deception, each with both beneficial and toxic personalities, so too does globalization carry this duality. Globalization has integrated developing countries into the global economy and allowed them to profit from new resources and advances. But globalization has also fragmented these countries by giving them access to resources that corrupt, such as arms for guerrilla leaders and rogue armies. What has changed, perhaps as early as the 1990s, is a new form of war, one that is tied to the signature of evil and its expression as gratuitous cruelty. No longer are wars confined to state borders, restricted to states and their legitimized militaries, financed by governments and tax revenues, and focused on combatants. Instead, the new wars of the twenty-first century have entirely porous boundaries, are funded by private organizations, run by grass-root groups, and motivated by the use of horrific means to achieve equally horrific ends, including torture, rape, mutilation, and the use and abuse of civilians, women, children and men alike. As a result, international law is effectively ineffective. Those running these new wars are outside of international law.

The consequence of the new wars extends beyond the travesties experienced by those living in these hot spots to the humanitarian aid workers and journalists who attempt to help the victims. Humanitarian aid is often pirated by rogue militias and journalists are frequently killed or badly injured. We must therefore face the sad reality that as we ended the twentieth-century and launched the twenty-first, casualties to non-combatant civilians shifted from few to many. We must face the reality that combating the potential for horrific atrocities will require new laws and new protections for those who risk their lives to aid victims and give voice to their often silent suffering.

We won't eradicate evil. Why? Because the capacity for evil is rooted in human nature, born of a creative mind that enables ideas and feelings to flip between beneficial and toxic. Though we institute programs and practices that promote the beneficial, living within every human mind is a toxic neighbor, waiting to move in. Adhering to authorities is beneficial in that great leaders are energizing, empowering, creative, and a source of guidance into a brighter future. But even great leaders can turn toxic, imposing corrosive ideologies and eliminating basic human rights. Conformity is beneficial in that we want to live in a society where norms are followed, providing stability and cooperation. But conformity is toxic when it leads to blind faith and uncritical thinking. Dehumanization is beneficial in allowing us to carry out medical procedures and live with certain kinds of human suffering. But dehumanization is toxic when it facilitates ethnic cleansing by shrinking the moral circle, turning atrocities into virtuous offerings. Tolerance and pluralism are beneficial in that they lead to respect and concern for others' attitudes and desires. But tolerance and pluralism are toxic when they breed apathy and a willingness to stand by as passive bystanders.

My diagnosis of evil is not meant to be defeatist, but realist. It is only through an acknowledgment of our biology and the environments it has created — and can create — that we can look for solutions to ameliorate the human condition. We are all vulnerable to walking on the wrong side. We are fallible. We are also enormously creative, capable of great change. Like no other species, we relentlessly seek novelty. No one wants to be like his or her predecessor. Whether it is a new culinary tradition, extreme sport, technological innovation, musical genre, or weapon of destruction, our search for novelty is an indestructible component of human nature. It is this capacity that is, in part, responsible for the general decline in violence since the beginning of human history, a point developed in great detail by Steven Pinker in his treatise *The Better Angels of Our Nature*. With our capacity to reason and to think of alternative ways of being, we have greatly reduced cruelty to humans and other animals. This is a good thing. But if my diagnosis of evil is correct, we must always be vigilant, as periods of peace can so easily be disrupted by violence, excessive, over-the-top violence.

Our journey into the nature of evil has come to an end. Bombarded by the sheer magnitude of lives lost or damaged beyond repair, it is natural to deaden our senses and choke our feelings in the hope of finding solitude and peace. As painful as a re-awakening is, we must remember the individuals that make up these massive atrocities. Reflecting upon the loss of his son who was murdered by the Lord's Resistance Army, an 80 year old Ugandan chief summed it up – "We have been forgotten. It's as if we don't exist."[64]

We must never forget. We must never deny our potential to cause horrific pain and suffering while finding ways to forgive and express deep compassion. We must never give up on humanity.

---

[64] *Human Rights Watch Report*, 2010, "The Trail of Death."

Recommended Books

Glover, J. (2000). *Humanity*. New Haven: Yale University Press.

Grossman, D. (1996). *On Killing*. New York: Back Bay Books.

Moore, M.S. (2010). *Placing Blame*. Oxford: Oxford University Press.

Pinker, S. (2011). *The Better Angels of Our Nature*. New York, Viking/Penguin.

www.ingramcontent.com/pod-product-compliance
Lightning Source LLC
Chambersburg PA
CBHW020728180526
45163CB00001B/159